Contemporary's

NUMBER POWER

Advanced

CB

CONTEMPORARY BOOKS

a division of NTC/CONTEMPORARY PUBLISHING GROUP
Lincolnwood, Illinois USA

ISBN: 0-8092-0612-9

Published by Contemporary Books,
a division of NTC/Contemporary Publishing Group, Inc.,
4255 West Touhy Avenue,
Lincolnwood (Chicago), Illinois 60646-1975 U.S.A.

890 CU 0987 654321

Contents

Contents

Contents

To the Learner

Even if math has never been easy for you, this text will give you the instruction and practice you need to understand the basics. In this global and technological society, an understanding of math is important, and at some time or another, you will be asked to demonstrate that you can solve math problems well.

Using *Number Power Advanced, Level A* is a good way to develop and improve your mathematical skills. It is a comprehensive text for mathematical instruction and practice. Beginning with a review of basic comprehension skills in addition, subtraction, multiplication, and division, *Number Power Advanced, Level A* includes basic concepts about performing basic computations. Then there is plenty of practice with decimals, fractions, signed numbers, and ratio and proportion.

Accompanying all of this practice are a Skills Inventory Pre-Test and a Skills Inventory Post-Test. The Skills Inventory Pre-Test will help you identify your math strengths and weaknesses before you begin working in the book. Then you can work in those areas where additional instruction and practice are needed. Upon completion of these exercises, you should take the Skills Inventory Post-Test to see if you have achieved mastery. Mastery is whatever score you and your instructor have agreed upon to be correct to insure that you understand each group of problems.

Usually mastery is completing about 80 percent of the problems correctly. After achieving mastery, you should then move to the next section of instruction and practice. In this way the text offers you the chance to learn at your own pace, covering only the material that you need to learn. In addition, the instruction on each page offers you the opportunity to work on your own.

Addition, subtraction, multiplication, division, decimals, fractions, signed numbers, ratio and percent instruction and practice are presented in the first part of *Number Power Advanced, Level A*. The other part contains applications. It includes numeration, number theory, data interpretation, algebra, measurement, and geometry. You will also work with ordinal numbers, place value of numbers, graphs, tables, charts, number sentences, calendars, time, plane figures, logical reasoning, problem solving, estimation, and many other topics.

Completing Number Power Advanced, Level A will make you more confident about doing mathematical problems. Remember to use the Answer Key in the back of the book to check your responses. Soon you will find yourself either enjoying math for the first time or liking it even more than you did previously.

Part A: Computation

Circle the letter for the correct answer to each problem.

1

$71.8 - 0.5 =$ _____

- **A** 21.8
- **B** 66.8
- **C** 713
- **D** 71.3
- **E** None of these

2

$x + x + x =$ _____

- **F** $3 + x$
- **G** x^3
- **H** $3x$
- **J** xxx
- **K** None of these

3

$3.1 \times 0.06 =$ _____

- **A** 186
- **B** 18.6
- **C** 1.86
- **D** 0.186
- **E** None of these

4

$$\begin{array}{r} \frac{1}{3} \\ -\frac{1}{6} \\ \hline \end{array}$$

- **F** $\frac{1}{6}$ **H** $\frac{1}{18}$
- **G** $\frac{2}{3}$ **J** $\frac{1}{2}$
- **K** None of these

5 $8[7 - (2 + 2)] =$ _____

- **A** 24   **C** 52
- **B** 56   **D** 42
- **E** None of these

6 15% of \$75.00 = _____

- **F** \$15.00
- **G** \$11.25
- **H** \$60.00
- **J** \$11.05
- **K** None of these

7

$-15 + (-5) =$ _____

- **A** 20
- **B** −20
- **C** 15
- **D** −15
- **E** None of these

8

$\frac{-30}{5} =$ _____

- **F** 6
- **G** 5
- **H** −6
- **J** −5
- **K** None of these

9

$\frac{2}{3} + \frac{1}{2} + \frac{1}{4} =$ _____

- **A** $\frac{1}{3}$ **C** $1\frac{1}{3}$
- **B** $1\frac{7}{12}$ **D** $1\frac{1}{2}$
- **E** None of these

10

14.5% of ☐ = 145

- **F** 1,000
- **G** 100
- **H** 10
- **J** 0.001
- **K** None of these

11

$0.09\overline{)4.68}$

- **A** 52
- **B** 5.2
- **C** 0.52
- **D** 0.052
- **E** None of these

12

$\frac{4}{5} \times 6\frac{2}{3} =$ _____

- **F** $6\frac{8}{15}$ **H** $4\frac{12}{15}$
- **G** $16\frac{8}{15}$ **J** $5\frac{1}{3}$
- **K** None of these

13

What percent of 360 is 30?

- **A** $83\frac{1}{3}\%$
- **B** 0.12%
- **C** 10.8%
- **D** 12%
- **E** None of these

14 $3x - y(x + 5) =$ _____

- **F** $\frac{x}{3} - yx + 5$
- **G** $2x - 5y$
- **H** $3x - xy - 5y$
- **J** $3x - y + x + 5$
- **K** None of these

Part B: Applied Mathematics

Circle the letter for the correct answer to each question.

15 Which group of fractions is in order from least to greatest?

  A $\frac{1}{2}, \frac{2}{3}, \frac{1}{4}, \frac{5}{6}$

  B $\frac{1}{4}, \frac{1}{2}, \frac{2}{3}, \frac{5}{6}$

  C $\frac{1}{4}, \frac{1}{2}, \frac{5}{6}, \frac{2}{3}$

  D $\frac{5}{6}, \frac{2}{3}, \frac{1}{2}, \frac{1}{4}$

16 If you are estimating by rounding to the nearest whole number, what numbers should you use to estimate $9.15 \div 1.7$?

  F 9 and 1
  G 8 and 1
  H 9 and 2
  J 8 and 2

17 In the number 190,567, what does the digit 9 represent?

  A 9 thousands
  B 9 ten-thousands
  C 9 hundred-thousands
  D 90 hundreds

18 Which of these is another way to write 61,000,000?

  F 6.1×10^7
  G 6.1×10^6
  H 0.61×10^7
  J 6.1×10^{-7}

19 Shirley bought 6 pounds of chocolates for $5.11 per pound. She divided them into 12 boxes, and she paid 75 cents per box to have them gift wrapped. Which of these number sentences could you use to find how much Shirley spent in all?

  A $(6 \times \$5.11) + (\frac{6}{12} \times \$0.75) = \square$

  B $6(\$5.11 + \$0.75) = \square$

  C $12(\$5.11 + \$0.75) = \square$

  D $(6 \times \$5.11) + (12 \times \$0.75) = \square$

20 There are sixty-two people in the Freedom Gospel Choir. Twenty-one of those people are men. What percent of the choir is female?

  F 33.3%
  G 50%
  H 66.1%
  J 45%

There are 40 million non-Christians living in North America. This graph shows the categories of their religious groups. Study the graph. Then do Numbers 21 through 25.

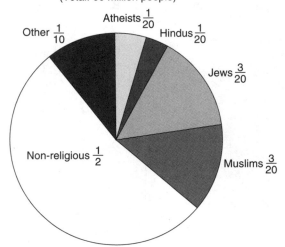

**Non-Christian Religious Groups
in North America Now**
(Total: 60 million people)

21 How many Hindus and Jews are there altogether in North America?

   A   8 million
   B   12 million
   C   2 million
   D   6 million

22 There are 255 million Christians living in North American. What is the ratio of Christians to non-Christians on the continent?

   F   8 : 51
   G 63 : 1
   H 43 : 8
   J   17 : 4

23 What percentage of non-Christians in North America are atheists?

   A 20%
   B 15%
   C 12%
   D   5%

24 Together, atheists, Hindus, and Muslims are what fraction of non-Christians in North America?

   F $\frac{1}{5}$ **H** $\frac{1}{3}$

   G $\frac{1}{4}$ **J** $\frac{2}{5}$

25

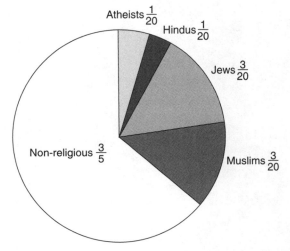

**Non-Christian Religious Groups
in North America in 1994**

This graph shows the non-Christian religious groups in 1994. Which group has grown the most since then?

   A non-religious
   B atheists
   C Muslims
   D other

The "Wind chill" is a measure of how cold it feels outside. It takes into account both the temperature and the wind speed. This table shows different wind chills. Temperatures are shown at the top of the table. Wind speeds are shown down the left side. Study the table. Then do Numbers 26 through 29.

Wind Chill

	30°F	20°F	10°F	0°F	−10°F	−20°F
5 mph	27	16	7	−5	−15	−26
10 mph	16	3	−9	−22	−34	−46
15 mph	9	−5	−18	−31	−45	−58
20 mph	4	−10	−24	−39	−53	−67
25 mph	1	−15	−29	−44	−59	−74
30 mph	−2	−18	−33	−49	−64	−79
35 mph	−4	−20	−35	−52	−67	−82
40 mph	−5	−21	−37	−53	−69	−84
45 mph	−6	−22	−38	−54	−70	−85

26 It is 20°F outside and the wind speeds rise from 10 miles per hour to 40 miles per hour. How much colder does it feel?

    F 3 degrees colder
    G 18 degrees colder
    H 21 degrees colder
    J 24 degrees colder

27 If the pattern of wind chills at 10 mph continues, what would be the wind chill at 10 mph and −30°F?

    A −51
    B −58
    C −62
    D −68

28 What temperature in degrees Celsius is equal to 30°F? *Hint:* You can use the conversion formula $C = \frac{5}{9}(F - 32)$.

    F about −1°C
    G about $34\frac{1}{2}$°C
    H about $-34\frac{1}{2}$°C
    J $-16\frac{2}{3}$°C

29 A teacher wants to show how wind chill changes as wind speeds rise. She makes a graph showing the wind chill at 10°F for different wind speeds. Which of these graphs correctly shows the relationship between wind chill and wind speed at 10°F?

A
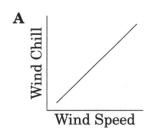

C
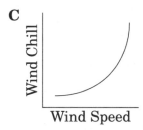

B
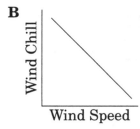

D
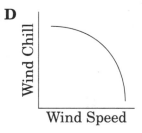

This map shows the first floor of a popular gym. Study the map. Then do Numbers 30 through 34.

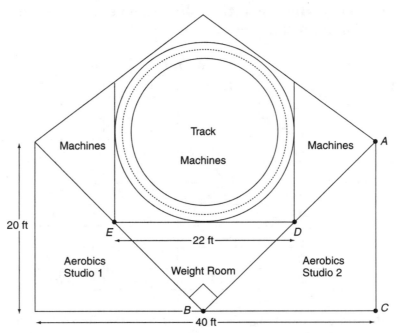

30 The two aerobics studios form congruent triangles. Therefore, the measure of $\angle ABC$ must be ? .

    F   30°
    G   40°
    H   45°
    J   50°

31 What type of polygon do the outside walls of the gym form?

    A   an irregular hexagon
    B   an irregular pentagon
    C   a regular pentagon
    D   a regular hexagon

32 What is the length of wall AB?

    F $\sqrt{800}$ feet
    G   200 feet
    H   250 feet
    J $\sqrt{400}$ feet

33 The weight room and Aerobics Studio 2 form similar triangles. What relationship must exist between the lengths of their walls?

    A $\dfrac{AB}{BC} = \dfrac{BD}{BE}$

    B $\dfrac{AB}{BC} = \dfrac{BE}{DE}$

    C $\dfrac{AB}{DE} = \dfrac{BC}{BE}$

    D $\dfrac{AB}{BC} = \dfrac{BD}{BE}$

34 What is the area of each aerobics studio?

    F   400 square feet
    G   800 square feet
    H   200 square feet
    J   This cannot be determined.

Read the passage and study the chart below. Then do Numbers 35 through 38.

Pablo just got a job writing estimates for Acme Basement Waterproofing. The company gave him this list of guidelines to use when calculating how much it will cost to waterproof a basement wall.

Excavate inside wall	$70 per foot
Waterproof outside wall	$10 per foot
Install drainage tile	$11 per foot
Install sump pump	$550

	Basement Depth		
	4 feet	**6 feet**	**8 feet**
Excavate outside wall	$47 per foot	$70 per foot	$91 per foot

35 Pablo's first customer has a basement that is 6 feet deep. She needs to have 11.5 feet of the outside wall excavated. Then the wall must be waterproofed and drainage tile must be installed. Which of the following is the best estimate of how much the entire job will cost?

   A $800.00
   B $850.00
   C $910.00
   D $700.00

36 Pablo's second customer has a basement that is 8 feet deep. He needs 39 feet of drainage tile installed. About how much can he save if he excavates the inside wall instead of excavating the outside wall?

   F $300.00
   G $800.00
   H $1,500.00
   J $2,000.00

37 Pablo's third customer says that her basement is $5\frac{1}{2}$ meters deep. Which of the following is the best estimate of her basement depth in feet? (Use the relationship that 1 meter is about 1.09 yards.)

   A   4 feet
   B   5 feet
   C   6 feet
   D   7 feet

38 The least expensive sump pump that Acme Waterproofing uses can pump up to 20 gallons of water per hour. How long would it take to drain a 1,000-gallon lake using that pump?

   F   8 hours, 20 minutes
   G   50 hours
   H   5 hours
   J   3 hours, 45 minutes

Skills Inventory Pre-Test Evaluation Chart

Use the key to check your answers on the Skills Inventory Pre-Test. The Evaluation Chart shows where you can turn in the book to find help with any problems you missed.

Key

1	D
2	H
3	D
4	F
5	A
6	G
7	B
8	H
9	E
10	F
11	A
12	J
13	E
14	H
15	B
16	H
17	B
18	F
19	D
20	H
21	B
22	J
23	D
24	G
25	D
26	J
27	B
28	F
29	D
30	H
31	B
32	F
33	C
34	H
35	C
36	G
37	C
38	G

Evaluation Chart

Problem Numbers	Skill Areas	Practice Pages
1, 3, 11	Decimals	20–31
4, 9, 12	Fractions	32–46
7, 8, 26	Integers	47–53
6, 10, 13, 20, 22	Ratios/Proportions/ Percent	54–64
15, 17, 18, 23	Numeration/Number Theory	1, 20–21, 32, 47, 78, 83, 88
24, 25, 27, 29	Data Interpretation	65–77
2, 5, 14, 19	Pre-Algebra/ Algebra	78–85, 86–104
34, 37, 38	Measurement	105–121
30, 31, 33	Geometry	122–136
21, 32, 36	Computation in Context	11–14, 29, 44, 51, 62, 70, 102, 119
16, 28, 35	Estimation/ Rounding	15–17

Correlations Between *Number Power Advanced and TABE*™ *Mathematics Computation*

Decimals   **Pre-Test Score** ☐   **Post-Test Score** ☐

Subskill	TABE, Form 7	TABE, Form 8	Practice and Instruction Pages in This Text (*p* means practice page)	Additional Practice and Instruction Resources
Addition	1	2	22, 30–31*p*	*Breakthroughs in Math/Bk. 2,* pages 34–43 *Number Power,* Bk. 2, pages 49–55 *GED Satellite Program: Mathematics,* pages 53–61
Subtraction	2	1	23, 30–31*p*	*Breakthroughs in Math/Bk. 2,* pages 34–41, 44–45 *Number Power,* Bk. 2, pages 49–54, 57 *GED Satellite Program: Mathematics,* pages 53–60, 61–62
Multiplication	10	8	24–25, 30–31*p*	*Breakthroughs in Math/Bk. 2,* pages 34–41, 48–51 *Number Power,* Bk. 2, pages 49–54, 59–62 *GED Satellite Program: Mathematics,* pages 53–60, 63–65
Division	15	6, 7	26–28, 30–31*p*	*Breakthroughs in Math/Bk. 2,* pages 34–41, 52–57 *Number Power,* Bk. 2, pages 49–54, 64–68 *GED Satellite Program: Mathematics,* pages 53–60, 66–68

Corresponds to TABE™ *Forms 7 and 8*
Tests of Adult Basic Education are published by CTB Macmillan/McGraw-Hill.
Such company has neither endorsed nor authorized this test preparation book.

Fractions   **Pre-Test Score** ☐   **Post-Test Score** ☐

Subskill	TABE, Form 7	TABE, Form 8	Practice and Instruction Pages in This Text (*p* means practice page)	Additional Practice and Instruction Resources
Addition	5	14	34–39, 45–46p	*Breakthroughs in Math/Bk. 2,* pages 66–75, 82–87, 104–105 *Number Power,* Bk. 2, pages 5–19 *GED Satellite Program: Mathematics,* pages 79–90
Subtraction	3, 6	5, 16	34–39, 45–46p	*Breakthroughs in Math/Bk. 2,* pages 66–72, 76–79, 82–87 *Number Power,* Bk. 2, pages 5–12, 17, 21–26 *GED Satellite Program: Mathematics,* pages 90–99
Multiplication	4	23	33–35, 40–42, 45–46p	*Breakthroughs in Math/Bk. 2,* pages 66–72, 92–95 *Number Power,* Bk. 2, pages 5–12, 28–32 *GED Satellite Program: Mathematics,* pages 100–106
Division	9	12	34–35, 43, 45–46p	*Breakthroughs in Math/Bk. 2,* pages 66–72, 96–99 *Number Power,* Bk. 2, pages 5–12, 34–41 *GED Satellite Program: Mathematics,* pages 106–110

Algebraic Operations   **Pre-Test Score** ☐   **Post-Test Score** ☐

Subskill	TABE, Form 7	TABE, Form 8	Practice and Instruction Pages in This Text (*p* means practice page)	Additional Practice and Instruction Resources
Algebra Operations	13, 18, 20, 21, 24	3, 10, 11, 18, 25	78–83, 84–85p, 93–100, 103–104p	*Real Numbers,* Bk. 5, pages 5–10, 12–15, 22–26, 52–59 *Number Power,* Bk. 3, pages 24–27, 45–53, 56–63, 66–69, 78, 100–123 *GED Satellite Program: Mathematics,* pages 199–201, 204–209, 211–212, 233–235, 292–293

Integers   **Pre-Test Score** ☐   **Post-Test Score** ☐

Subskill	TABE, Form 7	TABE, Form 8	Practice and Instruction Pages in This Text (*p* means practice page)	Additional Practice and Instruction Resources
Addition	17	9, 19	48, 52–53*p*	*Real Numbers,* Bk. 5, pages 30–36, 41, 49 *Number Power,* Bk. 3, pages 4–11, 18 *GED Satellite Program: Mathematics,* pages 279–283, 289
Subtraction	8, 25	17	49, 52–53*p*	*Real Numbers,* Bk. 5, pages 30–31, 35, 37–41, 49 *Number Power,* Bk. 3, pages 4–5, 12–13, 18 *GED Satellite Program: Mathematics,* pages 279–280, 283–285, 289
Multiplication	14	4	50, 52–53*p*	*Real Numbers,* Bk. 5, pages 30–31, 35, 43–45, 49 *Number Power,* Bk. 3, pages 4–5, 14–15, 19 *GED Satellite Program: Mathematics,* pages 279–280, 285–287, 289
Division	7, 11	13	50, 52–53*p*	*Real Numbers,* Bk. 5, pages 30–31, 35, 46–49 *Number Power,* Bk. 3, pages 4–5, 16–17 *GED Satellite Program: Mathematics,* pages 279–280, 287–289

Percent   **Pre-Test Score** ☐   **Post-Test Score** ☐

Subskill	TABE, Form 7	TABE, Form 8	Practice and Instruction Pages in This Text (*p* means practice page)	Additional Practice and Instruction Resources
Percents	12, 16, 19, 22, 23	15, 20, 21, 22, 24	57–61, 63–64*p*	*The GED Math Problem Solver,* pages 176–183, 186–193 *Number Power,* Bk. 2, pages 76–82, 85–86, 89–90 *GED Satellite Program: Mathematics,* pages 131–143

Correlations Between *Number Power Advanced and TABE*™ *Applied Mathematics*

Numeration   **Pre-Test Score** ☐   **Post-Test Score** ☐

Subskill	TABE, Form 7	TABE, Form 8	Practice and Instruction Pages in This Text (*p* means practice page)	Additional Practice and Instruction Resources
Recognizing Numbers	8	8	1–2, 18–19*p*, 20, 30–31*p*, 32–35, 45–46*p*, 47, 52–53*p*, 78	*Number Power Review,* pgs 2–3, 16–17, 44–47, 50–51, 58 *Breakthroughs in Math / Bk. 1,* pages 7–11, 14–15; *Bk. 2,* pages 30–38, 66–67 *GED Satellite Program: Mathematics,* pages 11–12, 53–57, 79–83, 279–280
Ordering	2	35	2, 18–19*p*, 21, 30–31*p*, 35, 45–46*p*, 47, 52–53*p*, 57	*Number Power Review,* pages 2–4, 16–17, 44–63 *Breakthroughs in Math / Bk. 2,* pages 30–39, 66–71, 106–107 *GED Satellite Program: Mathematics,* pages 53–58, 79–85, 111–114, 279–280
Place Value		1, 2	1–2, 16, 18–19*p*, 20–21, 30–31*p*	*Number Power Review,* pages 2–3, 46 *Breakthroughs in Math / Bk. 1,* pages 8–9; *Bk. 2,* pages 34–36 *GED Satellite Program: Mathematics,* pages 11–12, 53–54
Comparison		10	2, 21, 30–31*p*, 35, 45–46*p*, 47, 52–53*p*, 54, 57, 63–64*p*, 78, 82, 84–85*p*, 106, 109, 122	*Number Power Review,* pages 2–4, 16–17, 44–63 *Breakthroughs in Math / Bk. 2,* pages 30–39, 66–71, 106–107 *GED Satellite Program: Mathematics,* pages 53–58, 79–85, 111–114, 279–280
Fractional part	48		21, 32–35, 45–46*p*, 50, 52–53*p*, 68, 73, 105	*Number Power,* Bk. 2, pages 5–9 *Breakthroughs in Math / Bk. 2,* pages 66–69, 72 *GED Satellite Program: Mathematics,* pages 79–89
Scientific Notation	22, 50		81–82, 84–85*p*	*The GED Math Problem Solver,* pages 102–105 *GED Satellite Program: Mathematics,* pages 311–313
Roots		25	83, 84–85*p*	*The GED Math Problem Solver,* pages 88–95 *Number Power,* Bk. 3, pages 24–33 *GED Satellite Program: Mathematics,* pages 233–237, 296–297

Number Theory   **Pre-Test Score** ☐   **Post-Test Score** ☐

Subskill	TABE, Form 7	TABE, Form 8	Practice and Instruction Pages in This Text (*p* means practice page)	Additional Practice and Instruction Resources
Sequence	28, 43, 44		72, 86	*Number Power Review,* pages 98–99, 154–155 *GED Satellite Program: Mathematics,* pages 41–43, 186–190 *Critical Thinking with Math,* pages 4–5, 8–12, 58–59
Equivalent form	1		1–2, 20–22, 27–28, 33–35, 38–41, 43, 45–46p, 48–50, 52–53p, 54–58, 63–64p, 78, 81–83, 84–85p, 96–99, 103–104p, 107–110, 117–118p	*The GED Math Problem Solver,* pages 118–125, 176–178 *Number Power, Bk. 2,* pages 8–12, 49, 52–53, 76–80 *GED Satellite Program: Mathematics,* pages 83–89, 93–96, 111–112, 131–138
Multiples		20	38–39, 86	*Breakthroughs in Math / Bk. 2,* pages 82–85 *Critical Thinking with Math,* pages 3–7
Ratio, Proportion		41, 46	34, 54–56, 63–64p	*The GED Math Problem Solver,* pages 160–175 *Number Power, Bk. 6,* pages 81–96 *GED Satellite Program: Mathematics,* pages 121–129
Percents	9, 35	13, 14, 21	57–62, 63–64p, 69, 73, 75p	*The GED Math Problem Solver,* pages 176–183, 186–193 *Number Power, Bk. 2,* pages 76–95 *GED Satellite Program: Mathematics,* pages 131–163

Data Interpretation   **Pre-Test Score** ☐   **Post-Test Score** ☐

Subskill	TABE, Form 7	TABE, Form 8	Practice and Instruction Pages in This Text (*p* means practice page)	Additional Practice and Instruction Resources
Graphs	11, 12, 17, 18, 19, 49	44, 45	66–69, 71, 73–75, 76–77p	*Number Power, Bk. 8,* pages 1–32, 38–54, 72–99, 134–144, 156–159 *Real Numbers, Bk. 3,* pages 12–43, 61–66 *GED Satellite Program: Mathematics,* pages 177–187, 191–193
Tables, Charts, Diagrams	27, 36, 40, 42	11, 12, 17, 19, 23, 24	65, 70p, 72, 74, 76–77p	*Number Power, Bk. 5,* pages 67–91 *Real Numbers, Bk. 3,* pages 1–11, 44–50, 61–65 *GED Satellite Program: Mathematics,* pages 188–193

Algebra   **Pre-Test Score** ☐   **Post-Test Score** ☐

Subskill	TABE, Form 7	TABE, Form 8	Practice and Instruction Pages in This Text (*p* means practice page)	Additional Practice and Instruction Resources
Function/ Pattern		9, 34	72, 86	*Number Power Review,* pages 98–99, 154–155 *GED Satellite Program: Mathematics,* pages 41–43 *Critical Thinking with Math,* pages 4–5, 8–12
Missing Element		26, 28	55–56, 63–64*p*, 86–88, 103–104*p*	*The GED Math Problem Solver,* pages 18–19, 22–24, 68–70, 104–105 *Number Power Review,* pages 5, 172–175, 216–217 *GED Satellite Program: Mathematics,* pages 199–201, 64–65, 311–313
Number Sentence		3	11–14, 18–19*p*, 45–46*p*, 63–64*p*	*Number Power,* Bk. 7, pages 14–21, 26–43, 70–81 *GED Satellite Program: Mathematics,* pages 21–28, 37–39, 48–49 *Critical Thinking with Math,* pages 26–47
Equations	7, 10, 20, 38	18, 22	55–56, 63–64*p*, 87–99, 102, 103–104*p*	*Real Numbers,* Bk. 5, pages 1–29, 52–64 *Number Power,* Bk. 3, pages 34–77, 132–151 *GED Satellite Program: Mathematics,* pages 121–129, 155–159, 195–221
Inequality	4, 24		100, 103–104*p*	*GED Satellite Program: Mathematics,* pages 199–201, 206–209, 290–291

Measurement   **Pre-Test Score** ☐   **Post-Test Score** ☐

Subskill	TABE, Form 7	TABE, Form 8	Practice and Instruction Pages in This Text (*p* means practice page)	Additional Practice and Instruction Resources
Money	3		3, 16, 18–19p, 20, 25, 30–31p, 59	*The GED Math Problem Solver,* pages 12–14, 53, 68, 174 *Number Power Review,* pages 7, 22–25, 30–33, 46, 48, 60, 96–97 *Breakthroughs in Math / Bk. 1,* pages 14–16, 26–27, 37, 50, 64, 81, 91, 112
Time	5, 26	27	105–107, 110–111, 116, 117–118p	*Number Power, Bk. 9,* pages 126–156 *Breakthroughs in Math / Bk. 1,* pages 140–153 *GED Satellite Program: Mathematics,* pages 165–171
Temperature		4	47–48, 51–53p, 105–111, 117–118p	*Real Numbers, Bk. 4,* pages 46–53 *Number Power, Bk. 9,* pages 74–79 *GED Satellite Program: Mathematics,* pages 279–289
Length	25		105–113, 117–118p	*Number Power, Bk. 9,* pages 18–37 *Breakthroughs in Math / Bk. 1,* pages 140–153 *GED Satellite Program: Mathematics,* pages 165–173
Area	21, 34	42, 50	114, 117–118p	*The GED Math Problem Solver,* pages 61–65, 96–100 *Number Power, Bk. 4,* pages 84–85, 88–89, 92–93, 96–97, 100–101, 104–113 *GED Satellite Program: Mathematics,* pages 238–244, 248–251
Volume/ Capacity		15, 40	105–111, 115, 117–118p	*Number Power, Bk. 9,* pages 88–121 *Breakthroughs in Math / Bk. 1,* pages 140–153, 158–159 *GED Satellite Program: Mathematics,* pages 165–171, 245–251

Geometry   **Pre-Test Score** ☐   **Post-Test Score** ☐

Subskill	TABE, Form 7	TABE, Form 8	Practice and Instruction Pages in This Text (*p* means practice page)	Additional Practice and Instruction Resources
Geometric Elements	23	38	120–122, 132–133*p*	*Real Numbers,* Bk. 6, pages 1–15 *Number Power,* Book 4, pages 6–7, 22, 34–35 *GED Satellite Program, Mathematics,* pages 223–226
Plane Figures	16		112–114, 117–118*p*, 125–131, 132–133*p*	*Real Numbers,* Bk. 6, pages 1, 16–20 *Number Power,* Bk. 4, pages 82–83 *Pre-GED Satellite Program: Mathematics,* pages 117–120, 122–123
Coordinate Geometry		36	128–129	*The GED Math Problem Solver,* pages 40–42, 194–200 *Number Power,* Bk. 3, pages 84–96, 156–157 *GED Satellite Program: Mathematics,* pages 300–309
Visualization	13		119, 132–133*p*	*Real Numbers,* Bk. 6, pages 25, 34–35, 50–52
Angles	31	29, 31	122–125, 132–133*p*	*The GED Math Problem Solver,* pages 28–34 *Number Power,* Bk. 4, pages 6–25, 36–39 *GED Satellite Program: Mathematics,* pages 223, 252–257, 260–267
Similarity	15	32	129–130, 132–133*p*	*The GED Math Problem Solver,* pages 168–173 *Number Power,* Bk. 4, pages 44–53 *GED Satellite Program: Mathematics,* pages 260–264
Pythagorean Theorem	32	30, 39	128, 132–133*p*	*The GED Math Problem Solver,* pages 88–94 *Number Power,* Bk. 4, pages 56–61 *GED Satellite Program: Mathematics,* pages 268–270

Computation in Context   **Pre-Test Score** ☐   **Post-Test Score** ☐

Subskill	TABE, Form 7	TABE, Form 8	Practice and Instruction Pages in This Text (*p* means practice page)	Additional Practice and Instruction Resources
Whole Numbers	33	43	11–15, 17, 18–19*p*, 68, 70, 74–77*p*, 107, 111–116, 117–118*p*, 123, 128, 130–131, 132–133*p*	*The GED Math Problem Solver,* pages 2–17, 50–69, 76–87 *Number Power,* Bk. 6, pages 7–35, 55–69, 97–106, 125–131, 135–148 *GED Satellite Program,* pages 21–51
Decimals	14, 30, 41	33	23, 25, 27, 29, 30–31*p*, 74*p*, 113–115, 117–118*p*	*The GED Math Problem Solver,* pages 146–152 *Number Power,* Bk. 6, pages 36–47, 70–72, 107–113, 132–148 *GED Satellite Program,* pages 59–63, 69–77
Percent		47	57, 59–62, 63–64*p*, 69, 73, 75*p*	*The GED Math Problem Solver,* pages 176–193 *Number Power,* Bk. 6, pages 114–124, 132–148 *GED Satellite Program,* pages 144–160
Algebraic Operations	47	48	48–49, 51, 52–53*p*, 54–56, 63–64*p*, 82, 84–85*p*, 89–95, 101–102, 103–104*p*	*Real Numbers,* Bk. 5, pages 11, 16–21, 27–28, 50–51, 60–64 *Number Power,* Bk. 3, pages 41–42, 54–55, 64–65, 70–71, 73–74, 76–77, 132–151 *GED Satellite Program,* pages 126–128, 213–221, 297–299

Estimation   **Pre-Test Score** ☐   **Post-Test Score** ☐

Subskill	TABE, Form 7	TABE, Form 8	Practice and Instruction Pages in This Text (*p* means practice page)	Additional Practice and Instruction Resources
Reasonableness of Answer		6, 16, 49	15, 18–19*p*	*Critical Thinking with Math,* pages 50–51 *Number Power,* Bk. 7, pages 132–134 *GED Satellite Program,* pages 29–30
Rounding	37, 39	5, 7	16–17, 18–19*p*, 20, 25, 28, 30–31*p*, 61	*The GED Math Problem Solver,* pages 8, 12–13, 56–58, 132–133, 140–141, 186–187 *Number Power Review,* pages 6–7, 68–71, 104–107 *GED Satellite Program,* pages 30–33, 70
Estimation	6, 29, 45, 46	37	15–17, 18–19p, 30–31*p*, 66–67, 83, 105–106, 108, 122	*The GED Math Problem Solver,* pages 8, 12–13, 56–58, 132–133, 140–141, 186–187 *Number Power Review,* pages 8–13, 68–71, 104–107, 146–147 *GED Satellite Program,* pages 30–35

Whole Numbers

Place Value

> The ten **digits** are 0, 1, 2, 3, 4, 5, 6, 7, 8, and 9. The value of a digit in a number depends on its **place** in that number.

Look at the numbers 904 and 409. They have the same digits, but they are different numbers. That is because the digits are in different places. The number 904 stands for

900 + 4 *or* **9 hundreds and 4 ones**

The number 409 stands for

400 + 9 *or* **4 hundreds and 9 ones**

The diagram at the right shows the first seven place values for whole numbers. The number 815,203 has digits in the first six places. It has 8 hundred-thousands, 1 ten-thousand, 5 thousands, 2 hundreds, 0 tens, and 3 ones. It has no digit in the millions place.

	millions	hundred thousands	ten thousands	thousands	hundreds	tens	ones
	__ ,	8	1	5 ,	2	0	3

PRACTICE

Fill in the blanks on pages 1 and 2.

1 503,700 has

    5 _____ ,

    3 _____ ,

    7 _____ .

2 7,019,000 has

    7 _____ ,

    1 _____ ,

    9 _____ .

3 Look at the number 1,760,425. What is the place value of the digit 4? _____

4 Look at the number 513,402. What is the place value of the digit 5? _____

5 In the number 705,312, what digit is in the thousands place? _____

6 In the number 1,235,176, what digit is in the tens place? _____

7 In 50,321, the value of the digit 3 is three hundred. The value of the digit 5 is fifty thousand. What is the value of the digit 2? _____

8 What is the value of the digit 7 in the number 75,000? _____

9 What is the value of the digit 6 in the number 612,000? _____

10 What is the value of the digit 4 in the number 1,450? _____

11 What is the value of the digit 8 in the number 183,679? _____

When you say a number aloud, you group together the digits in the hundred-thousands place, ten-thousands place, and thousands place. Similarly, the digits in the hundred millions, ten millions, and millions places are grouped together.

$$41,032 = 41,000 + 32$$
$$= 41 \text{ thousand, thirty-two}$$

$$234,502,070 = 234,000,000 + 502,000 + 70$$
$$= \text{two hundred thirty-four million, five hundred two thousand, seventy}$$

Write each number below in words.

12 302 _____

_____

13 1,040 _____

_____

14 3,100 _____

_____

15 1,350,000 _____

_____

For problems 16–20, write each number in digits. Watch for number names that skip places. You must put a zero in that position.

16 twelve thousand, forty _____

17 fifty-nine thousand, six _____

18 three hundred twelve thousand _____

19 one thousand, one hundred two _____

20 ten million, five hundred thousand _____

21 If you change the digit 3 in the number 703,000 to a 4, how much does the value of the number increase? _____

22 If you change the digit 4 in the number 10,415 to a 6, how much does the value of the number increase? _____

23 If you change the digit 3 in the number 1,304,512 to a 1, how much does the value of the number change? _____

If two whole numbers have different numbers of digits, the number with more digits is larger.
    5 , 0 1 3   is larger than   9 8 4

If two whole numbers have the same number of digits, start at the left to compare them.
    5 , 8 6 2   is larger than   2 , 9 7 4
because 5 is larger than 2.

---

24 Circle the larger number.   7,100   1,700

25 Circle the larger number.   170,000   71,000

26 Circle the larger number.   90,000   100,000

27 Circle the larger number.   14,000   23,000

28 Circle the larger number.   10,399   10,605

29 Arrange the digits 6, 7, 0, and 5 to make the *largest possible* whole number.

_____

30 Arrange the digits 4, 6, 7, 1, and 9 to make the *smallest possible* whole number.

_____

Place Value

Review of Whole Number Addition

To add large numbers, start by writing them in column form. Start with the two digits in the right column. Add those digits, and write the sum. Then move one column to the left and add the digits in that column. Keep this up until you have added all the columns.

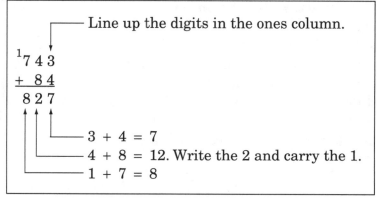

To check an answer in an addition problem, use subtraction to reverse or "undo" the addition. Subtract the bottom number in the problem from your answer. The result for the subtraction problem should be the same as the top number of the addition problem.

To check the problem above, subtract 84 from 827. The difference in 743, so the sum 827 is correct.

PRACTICE

Write each problem below in column form. Solve the problem and check your work. If a label such as a dollar sign or a unit name appears in a problem, repeat it in your answer.

1   76 + 15 = _____

2   165 + 307 = _____

3   1,130 + 350 = _____

4   25 gal + 98 gal = _____

5   341 liters + 92 liters = _____

6   5,232 + 309 = _____

7   8,205 + 4,520 = _____

8   211 lb + 35 lb + 12 lb = _____

9   61 + 103 + 59 = _____

10   823 + 256 + 301 = _____

11   1,356 mi + 2,459 mi = _____

Review of Whole Number Subtraction

To subtract, write the numbers in column form. The number you are subtracting *from* must be the top number. Starting at the right, subtract the digits in each column. If the difference in a column is zero, be sure to write a zero below the column. However, you *do not* write a zero at the left of a whole number.

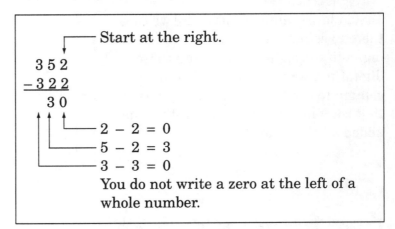

Start at the right.

$$352$$
$$-322$$
$$30$$

2 − 2 = 0
5 − 2 = 3
3 − 3 = 0

You do not write a zero at the left of a whole number.

Sometimes a digit in the bottom number is larger than the digit above it. Each time you "borrow," you replace the number you are *borrowing from* with a digit that is one less than that digit. Then you replace the digit that is too small with a number that is ten more. Be sure to write neatly.

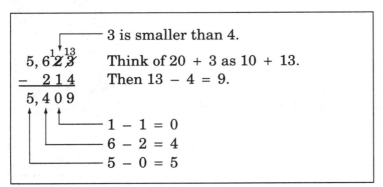

3 is smaller than 4.
Think of 20 + 3 as 10 + 13.
Then 13 − 4 = 9.

$$5,6\overset{1}{\cancel{2}}\overset{13}{\cancel{3}}$$
$$-214$$
$$5,409$$

1 − 1 = 0
6 − 2 = 4
5 − 0 = 5

To check this result, add the two bottom numbers, 214 and 5,409. The sum is the same as the top number 5,623. That means the original subtraction is correct.

PRACTICE

Write each subtraction problem in column form. Solve the problem and then use addition to check your work. If there is a label in a problem, make sure you include it in your answer.

1 98 mi − 56 mi = _____

2 311 − 290 = _____

3 980 − 190 = _____

4 36 in. − 13 in. = _____

5 157 − 25 = _____

6 817 − 326 = _____

7 92 − 25 = _____

8 1,230 ft − 18 ft = _____

9 73 − 58 = _____

In the following example, you must borrow from more than one column. *Remember:* Once you have crossed out a number and written a new number above it, you must use that new number.

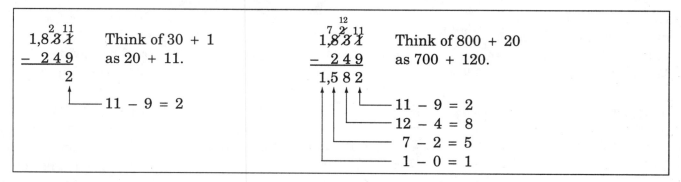

Here is an example that shows how to borrow from a string of zeros. In the ones place, 7 is less than 9. Starting at the digit 7, move to the left until you find the first nonzero digit.

$$\begin{array}{r} {}^{7\ 9\ 9\ 9\ 17} \\ 8\ 0,0\ 0\ 7 \\ -\ \ \ 1,0\ 1\ 9 \\ \hline 7\ 8,9\ 8\ 8 \end{array}$$

Write a digit one less than the first nonzero digit. Then add ten to the digit that is too small and write 9 over each digit in between.

PRACTICE

Solve each subtraction problem. Check your work.

10
$$\begin{array}{r} 452 \\ -\ 165 \\ \hline \end{array}$$

11
$$\begin{array}{r} 153 \text{ kilometers} \\ -\ 75 \text{ kilometers} \\ \hline \end{array}$$

12
$$\begin{array}{r} 612 \\ -\ 39 \\ \hline \end{array}$$

13
$$\begin{array}{r} 322 \text{ grams} \\ -155 \text{ grams} \\ \hline \end{array}$$

14
$$\begin{array}{r} 6,280 \\ -3,525 \\ \hline \end{array}$$

15
$$\begin{array}{r} 3,000 \\ -\ 209 \\ \hline \end{array}$$

16
$$\begin{array}{r} 10,000 \\ -\ 299 \\ \hline \end{array}$$

17
$$\begin{array}{r} 100 \text{ dollars} \\ -\ 27 \text{ dollars} \\ \hline \end{array}$$

18   450 − 99 = _____

19   400 − 339 = _____

20   212 − 39 = _____

21   5,100 − 149 = _____

Review of Whole Number Multiplication

Here is a multiplication problem in column form. Multiply the bottom number times each digit in the top number, working from right to left.

$$
\begin{array}{r}
{}^{2}3\,6\,2 \\
\times \quad 4 \\
\hline
1{,}4\,4\,8
\end{array}
$$

← Multiply by this digit.

4 × 2 = 8
4 × 6 = 24.
Write the 4 and carry the 2.
4 × 3 = 12, and then 12 + 2 = 14.
Write 14.

To check this multiplication problem, divide 1,448 by 4. The quotient is 362, which is the top number in the multiplication problem. The answer is correct.

PRACTICE

Write each multiplication problem in column form. Then find the product and check your work. *Hint:* **Write a digit in the product for every digit in the top number, even if the digit in the top number is a zero.**

1   65 × 6 = _____

2   4 × 150 = _____

3   5 × 250 kilometers = _____

4   21 feet × 7 = _____

5   303 × 6 = _____

6   1,321 × 2 = _____

7   170 × 2 = _____

8   6,100 × 3 = _____

9   510 × 8 = _____

In the following example, you are multiplying by a two-digit number. Always erase the carry digits after your first multiplication. That gives you room to write carry digits for the second multiplication.

Problem:	Step 1: Multiply by 5.	Steps 2 and 3: Start writing the product in the tens column, leaving a blank in the ones column. Multiply by 2, then add.
$\begin{array}{r} 56 \\ \times\ 25 \\ \hline \end{array}$	$\begin{array}{r} {}^{3}56 \\ \times\ 25 \\ \hline 280 \end{array}$	$\begin{array}{r} {}^{1}56 \\ \times\ 25 \\ \hline 280 \\ 112 \\ \hline 1{,}400 \end{array}$

PRACTICE

Write each multiplication problem in column form and then solve the problem. Check your work. *Remember:* Multiply *before* you add the digit being carried.

10 $152 \times 22 =$ _____

11 36 inches $\times$ 12 = _____

12 $190 \times 25 =$ _____

13 $2{,}119 \times 42 =$ _____

14 $250 \times 10 =$ _____

15 $424 \times 15 =$ _____

16 28 feet $\times$ 30 = _____

17 $346 \times 52 =$ _____

To multiply by a 3-digit number, multiply the top number by each digit in the bottom number. When you write the product for the third digit, leave the ones place and tens place blank and start writing in the hundreds place.

18 $113 \times 123 =$ _____

19 $432 \times 176 =$ _____

Review of Whole Number Division

To write a division problem using the bracket $\overline{\smash{)}}$, the number you are *dividing up* (the **dividend**) goes inside the bracket. The number you are *dividing by* (the **divisor**) goes to the left of the bracket. The answer in a division problem is called the **quotient.**

After you write the first digit in the quotient, use a zero as a placeholder whenever a digit in the dividend (inside the bracket) is smaller than the divisor.

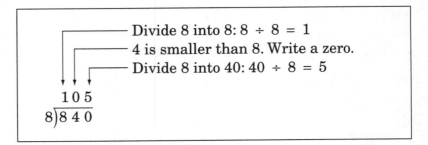

Divide 8 into 8: $8 \div 8 = 1$
4 is smaller than 8. Write a zero.
Divide 8 into 40: $40 \div 8 = 5$

$$\begin{array}{r} 1\,0\,5 \\ 8\overline{)8\,4\,0} \end{array}$$

To solve a problem with a remainder, you must find the largest whole number that goes into the dividend. The next step is to multiply that digit by the divisor. The last step is to subtract.

$$\begin{array}{r} 58 \text{ r } 3 \\ 4\overline{)235} \\ \underline{20} \\ 35 \\ \underline{32} \\ 3 \end{array}$$

1 Divide 4 into 23. Write 5 in the quotient.
2 Multiply: $5 \times 4 = 20$. Write 20.
3 Subtract: $23 - 20 = 3$. Bring down the next digit 5.
4 Divide 4 into 35. Write 8 in the quotient and then multiply: $8 \times 4 = 32$.
5 Subtract: $35 - 32 = 3$.

The quotient is 58 r 3.

To check the result of a division problem, multiply the whole-number part of the quotient and the divisor. Then add the remainder (if there is one). The result should be the number inside the division bracket. The check at the right is for the division problem $4\overline{)235}$.

$$58 \times 4 = 232$$
$$232 + 3 = 235$$

PRACTICE

Solve each problem and use multiplication to check your answer. *Hint:* **Do not skip any zeros in your answers.**

1 $3\overline{)639}$

2 $5\overline{)21}$

3 $6\overline{)366}$ miles

4 $5\overline{)255}$ gallons

5 $8\overline{)44}$

6 What is 3,216 divided by 4?

7 What is 51 divided by 7?

8 What is 510 divided by 5?

9 What is 71,428 divided by 7?

10 What is 50,010 divided by 5?

A division problem can become complicated. You should write each step neatly.

Look at the problem below. Be sure to notice that after the first three steps, you always follow the same four-step pattern:

- Bring down the next digit.
- Divide.
- Multiply.
- Subtract.

This sequence of steps is called **long division.**

$$
\begin{array}{r}
345 \\
4\overline{)1380} \\
\underline{12} \\
18 \\
\underline{16} \\
20 \\
\underline{20} \\
0
\end{array}
$$

1 Divide 4 into 13. Write 3.	**4** Bring down the next digit 8.	**8** Bring down the next digit 0.
2 Multiply: $3 \times 4 = 12$.	**5** Divide 4 into 18. Write 4.	**9** Divide 4 into 20. Write 5.
3 Subtract: $13 - 12 = 1$.	**6** Multiply: $4 \times 4 = 16$.	**10** Multiply: $5 \times 4 = 20$.
	7 Subtract: $18 - 16 = 2$.	**11** Subtract: $20 - 20 = 0$.

PRACTICE

Use long division to solve these problems. Show all your work and check your answers using multiplication.

11   $3\overline{)1071}$ 14 $3\overline{)264}$ liters   17 $5\overline{)215}$ meters

12   $4\overline{)920}$ hours   15 $7\overline{)371}$ 18 $6\overline{)3384}$

13   $5\overline{)835}$ 16 $4\overline{)9144}$ 19 $3\overline{)10182}$

Dividing by a Two-Digit Divisor

When you divide by a two-digit number, the first step is to find the digit for the quotient. Look at these two examples.

Example 1:

$$15\overline{)346}$$

quotient 23

$$
\begin{array}{r}
23 \\
15\overline{)346} \\
\underline{30} \\
46 \\
\underline{45} \\
1
\end{array}
$$

1 Divide 34 by 15. Write 2.
2 Multiply: $2 \times 15 = 30$.
3 Subtract: $34 - 30 = 4$.

4 Bring down the next digit 6.
5 Divide 46 by 15. Write 3.
6 Multiply: $3 \times 15 = 45$.
7 Subtract: $46 - 45 = 1$.

The quotient is 23 r 1.

Example 2:

$$
\begin{array}{r}
13 \\
23\overline{)299} \\
\underline{23} \leftarrow 1 \times 23 \\
69 \\
\underline{69} \leftarrow 3 \times 23 \\
0
\end{array}
$$

Important: The results of the "multiply" step *must not* be larger than the number above it. If it *is* larger, do the step again using a smaller digit in the divisor.

PRACTICE

Solve each problem below. Then check your work.

1 $22\overline{)245}$

2 $51\overline{)1320}$

3 $16\overline{)3240}$

4 $21\overline{)4862}$

5 $32\overline{)739}$

6 $23\overline{)269}$

7 $35\overline{)789}$

8 $50\overline{)13500}$

9 $32\overline{)6780}$

Solving Word Problems: The First Step

> When you are solving a word problem, the first step is to decide whether you should add, multiply, subtract, or divide.

Here are some clues to help you decide which operation to use:

- If you put different amounts together, you will add.
- If you compare one amount to another amount, you will subtract or divide.
- If you take something away from something else, you will subtract.
- If there are several of something, you will multiply or divide.

Here are some "signal words" for the four math operations of addition, subtraction, multiplication, and division.

Addition	Subtraction	Multiplication	Division
plus	minus	times	divide
sum	difference	multiplied by	apiece
total	take away	product	each
added to	subtract	twice, three	per
altogether	left over	times, and so on	
in all	how much	apiece	
combined	change	each	
increased by	more than		
	less than		
	decreased by		

PRACTICE

For each problem, tell whether you will add, subtract, multiply, or divide. You do not have to solve the problem.

1   You buy 128 dollars in groceries and write a check for 140 dollars. How much change should you get back?   _____

2   There will be 26 performances of *The Nutcracker*. About 450 people are expected to attend each performance. About how many people will attend in all?   _____

3   You buy a television for 425 dollars. Tax is 42 dollars. How much do you pay in all?   _____

4   Your family used 1,892 gallons of water in September. How many gallons of water did they use per day?   _____

5   If you family uses 1,892 gallons of water per month, how much will they use in a year?   _____

6   You need empty egg cartons for a craft project. You save 14 egg cartons. A neighbor gives you 7 cartons and your family collects 25 cartons. How many is that in all?   _____

Each problem below needs two steps to be solved. Circle the letter for the choice that describes how to solve the problem.

7 At Central Middle School there are 556 seventh graders, 397 eighth graders, and 35 teachers. How many teachers are there per student?

   A Add 556 and 397. Then divide the sum by 35.

   B Add 556 and 397. Then subtract 35 from the sum.

   C Add 556 and 397. Then add 35.

8 You need 60 ounces of nuts to make holiday candies. You buy 15 ounces of pecans, 12 ounces of walnuts, and 20 ounces of peanuts. How many more ounces of nuts do you need to buy?

   F Add 15, 12, and 20. Then multiply the sum by 60.

   G Add 15, 12, and 20. Then add the sum to 60.

   H Add 15, 12, and 20. Then subtract the sum from 60.

9 You have six 20-dollar bills in your wallet. You spend $15.96. How much cash do you have left?

   A Multiply 20 dollars by six. Then subtract $15.96.

   B Add 20 dollars and six. Then subtract $15.96.

   C Subtract $15.95 from $20.00. Then multiply by six.

10 John's daughter eats two apples a day. How many dozen apples does she eat in a year? (There are 365 days in a year and there are 12 apples in a dozen.)

   F Add 365 to 12. Then divide by 2.

   G Multiply 2 by 365. Then divide by 12.

   H Divide 365 by 2. Then divide again by 12.

11 Derek's living room is 150 square feet and his family room is 300 square feet. If he buys 500 square feet of carpeting for these two rooms, how much carpet will be left over?

   A Subtract 150 from 500. Then add 300 to the difference.

   B Add 150 and 300. Then divide 500 by the sum.

   C Subtract 150 from 500. Then subtract 300 from the difference.

12 Ellen's backyard is 45 feet wide and 25 feet deep. How much fencing would she need to enclose all four sides of the yard?

   F Add 45 and 25. Then divide the sum by 4.

   G Add 45 and 25. Then multiply the sum by 2.

   H Subtract 25 from 45. Then multiply the difference by 4.

13 Mike must move an oversized trailer 180 miles to California. He completes the trip in three days, driving 8 hours on Monday, 11 hours on Wednesday, and 10 hours on Thursday. On average, how many miles does he drive each hour?

   A Add 8 hours, 11 hours, and 10 hours. Then subtract 180miles from the sum.

   B Add 8 hours, 11 hours, and 10 hours. Then multiply the sum by 180 miles.

   C Add 8 hours, 11 hours, and 10 hours. Then divide 180 miles by the sum.

Solving Word Problems: The First Step

Solving Word Problems: The Next Steps

> The second step in solving a word problem is to find the numbers you will use to solve the problem.

When you solve real problems, often you do not have enough information or you have information that you do not need. Part of solving the problem is to decide what information is important and figure out how to get that information.

PRACTICE

The word problems below contain more information than you need. For each problem, circle the numbers you need to solve each problem.

1   Theo had to drive 840 miles in one day. He drove $3\frac{1}{2}$ hours and then took a 20-minute break. He drove another 5 hours and then took $1\frac{1}{2}$ hours for lunch. Then he drove 6 hours to finish the trip. How many hours did Theo spend driving?

2   Mark's company is getting $3,500 to do some of the carpentry work on a new house. Mark is trying to figure out how much it would cost him to have 2 master carpenters do all the work. He pays master carpenters $250.00 a day while apprentices only get $96.00. Two master carpenters could finish the job in 3 days. One master carpenter with 3 apprentices would take 5 days. If Mark puts two master carpenters on the job, how much will be pay?

3   There are six girls in the Maybury School Girl Scout troop. For an upcoming project, they each need a Styrofoam ball (which costs 45 cents), a bottle of paint (which costs 75 cents), a paintbrush (89 cents), and 6 ounces of beads ($0.25 per ounce). How much money will the Girl Scout troop spend altogether on beads?

4   Poinsettias are $15.00 each. Ryan needs enough poinsettias to go across a stage 35 feet wide. Each poinsettia is about 2 feet wide and weighs about $\frac{1}{2}$ pound. How many poinsettias does Ryan need?

The following word problems do not give enough information. On the blank lines, tell what else you would need to know before you could solve each problem.

5   The Darouie family's heating bill was $250 in December. How many gallons of heating oil did they use?

You also need to know _____

_____

6   About how much gas would Anne's car use on a 300-mile trip?

You also need to know _____

_____

The third step in solving a word problem is to write numbers and operation signs ($+$, $-$, $\div$, or $\times$). That is called **setting up the problem** or **writing a set-up** for the problem.

In a subtraction problem or in a division problem, the order of the numbers is important. For subtraction, the number you are *subtracting from* goes on top. For division, the number you are *dividing by* goes outside the bracket.

PRACTICE

For each problem below, set up the problem and then solve it.

7 There are 25 houses on each square block of the Shady Hills subdivision. There are 15 square blocks altogether. How many houses are there in the subdivision?

8 There are 16,420 unemployed adults in Marion. If the perfume factory closes, another 1,567 adults will lose their jobs. How many adults would be out of work if the perfume factory closes?

9 Eliza Holmes died in 1925 at the age of 95. In what year was she born?

10 A homeowner has 12 ounces of fertilizer. She has 120 square feet of lawn. How much fertilizer can she spread on each square foot of lawn?

The problems below are two-step or three-step problems. Take special care in setting them up.

11 For the first three months of last year, Jackie made $2,500 a month. Then she got a raise and made $3,000 a month. How much did Jackie make last year?

12 Lee spent 18 hours making a quilt. She spent 84 dollars on materials and she sold the quilt for 300 dollars. What was her profit in terms of dollars per hour?

13 The drive from Wilber to Crete is 25 miles. The drive from Crete to Lincoln is 65 miles. If you drive 45 miles an hour, how long will it take to drive from Wilber to Lincoln by way of Crete?

Estimating

Sometimes you need to know *about how much money* something costs or *about how much time* something takes. This kind of amount is called an **estimate.** An estimate is a number that is close to an actual or exact amount.

All the words in the box signal that a math problem calls for estimation.

> *almost*
> *approximately*
> *about*

PRACTICE

Circle the problem number for each situation below that calls for estimating.

1   The newspaper says that there are about 30 businesses downtown, but that figure seems too high to you. You decide to find out the number of businesses for yourself.

2   You just wrote a check for $65.00. You must subtract that amount from your old balance to find your current balance.

3   You want to know about how many fish are in the Great Lakes.

4   You want to know the approximate number of people who will be attending an annual pancake breakfast.

5   You need 35 milliliters of medicine. You want to know how many teaspoons that is.

6   You want to know the amount of money you spend each year at restaurants.

7   You sell someone a book for $18.42. He gives you a 20-dollar bill, and you need to know how much change to give back.

Often you can use common sense to come up with an estimate. For instance, you know from experience that a hamburger and an order of fries should cost about five dollars, not fifty dollars. This type of common sense is very useful when you are doing math problems. *Get into the habit of using it to check whether your answers make sense.*

8   The cost of 10 gallons of gasoline would be about __?__ .

   A $5.00   **C** $12.00
   B $1.50   **D** $20.00

9   An 800-mile drive on the highway would take about __?__ .

   F 12 hours   **H** 3 hours
   G 7 hours   **J** 3 days

10   If you earn $25,000 a year, in one month you would earn about __?__ .

   A $300   **C** $10,000
   B $2,000   **D** $800

11   Three ice cream cones would cost about __?__ .

   A 25 cents   **C** $5.00
   B 95 cents   **D** $12.00

12   An adult cat weighs about __?__ .

   F 3 lb   **H** 50 lb
   G 11 lb   **J** 75 lb

Rounding

Another way to estimate the answer to a math problem is to figure it out using rounded numbers. A rounded number is close to the exact one, but it is easier to work with.

To round a number, think of it as being on a hilly number line, as shown below. Each low spot ends in a zero. Numbers ending in 0, 1, 2, 3, and 4 roll back to the nearest low spot. Numbers ending in 5, 6, 7, 8, and 9 roll ahead to the nearest low spot.

You can round a number to different place values.

- 2,194 rounded to the tens place is 2,190.
- 2,194 rounded to the hundreds place is 2,200.
- 2,194 rounded to the thousands place is 2,000.

To round a number to a place value, look at the digit just to the right of that place value. If that digit is less than 5, round down. If that digit is 5 or more, round up.

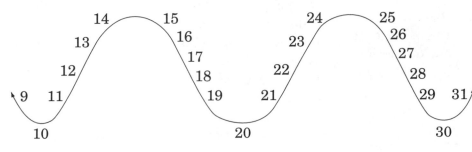

Examples:

11 rounds to 10.
18 rounds to 20.
20 rounds to 20.
5 rounds to 10.

PRACTICE

Circle the number that completes each statement.

1	102 rounds to	100	110
2	397 rounds to	390	400
3	158 rounds to	150	160
4	$79.36 rounds to	$79.30	$79.40
5	$112 rounds to	$110	$120
6	5,999 rounds to	6,000	5,990

Round each number to the indicated place.

7 679 to the hundreds place _____

8 1,723 to the hundreds place _____

9 1,723 to the thousands place _____

10 12,314 to the thousands place _____

11 $45,15 to the nearest dollar _____

12 $143.15 to the nearest ten dollars _____

13 $57.67 to the nearest ten dollars _____

14 56,097 to the hundreds place _____

15 1,352 to the tens place _____

16 2,425,980 to the thousands place _____

17 998 to the thousands place _____

18 $18.15 to the nearest dollar _____

19 $11.35 to the nearest dime _____

20 $112.45 to the nearest ten dollars _____

Rounding To Solve Word Problems

In a word problem, you can use the words *about, approximately,* or *almost* as a signal that the word problem calls for estimation. For this type of problem, set it up as you would any other word problem. Then round all the numbers before you do any figuring.

PRACTICE

Use this map for questions 1 through 10. Some of these problems are one-step problems. Some of the problems are two-step problems.

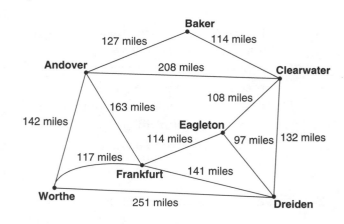

For problems 1–4 below, set up each problem. You do not have to solve it.

1   About how long is the trip from Clearwater to Worthe via Dreiden?

2   Clearwater is about how much farther from Andover than Frankfurt is?

3   About how long is the trip from Dreiden to Frankfurt if you go by way of Eagleton?

4   Three times a month, Shu makes a roundtrip drive from Andover to Worthe. Approximately how many miles is that?

Use estimation to solve each of the following problems. Round all numbers to the nearest ten.

5   Richard must drive from Dreiden to Clearwater, from there to Frankfort, and then back home to Dreiden. About how many miles will he drive?

6   Frank's car gets 28 miles to the gallon on the highway. About how much gas will he use on a round-trip drive between Frankfurt and Worthe?

7   Approximately how much farther is the trip from Dreiden to Andover if you go through Clearwater than if you go through Frankfurt?

Whole Numbers Skills Practice

Circle the letter for the best answer to each problem.

1   For which of these situations would you need an exact number?

   A   reporting your office's total expenses last month
   B   figuring how many sheets of paper your office uses in a year
   C   reporting how many calls the reception desk answered last year
   D   determining how many hours the workers in your office spend each year waiting to use the copy machine

2   You are estimating by rounding to the nearest 100. What numbers should you use to estimate 12,314 times 5,096?

   F   12,000 and 5,000
   G   12,300 and 5,000
   H   12,300 and 5,100
   J   12,310 and 5,100

3   There were 243 workdays last year, and Raul spent 43 minutes a day commuting to and from work. Which of these number sentences would give Raul the best estimate of how many hours he spent commuting last year? (*Remember:* There are 60 minutes in an hour.)

   A $(200 \times 40) \div 60 = \square$
   B $(240 \times 40) \div 60 = \square$
   C $(250 \times 40) \div 60 = \square$
   D $(240 \times 50) \div 60 = \square$

4   The newspaper reported that there are 31,000,000 households with dogs in the U.S. and 27,000,000 households with cats. To what place did they round these numbers?

   F   thousands
   G   hundred thousands
   H   millions
   J   ten millions

5   There are 314 students at Taylor Grade School, and each of them drinks 1 pint of milk a day. Which of these is the best estimate of how many pints of milk the school uses in seven school days?

   A   2,100
   B   2,800
   C   2,500
   D   2,400

6   Reginald's living room is 211 square feet, his family room is 359 square feet, and his office is 145 square feet. If it takes him 10 minutes to clean 100 square feet of carpet, about how long will it take him to clean the carpet in all three rooms?

   F   70 minutes (or 1 hour, 10 minutes)
   G   50 minutes
   H   90 minutes (or $1\frac{1}{2}$ hours)
   J   100 minutes (or 1 hour, 40 minutes)

7   One cookie contains 43 calories. Which of these is the best estimate of how many calories a box of 78 cookies would have?

   A   4,000
   B   3,200
   C   3,600
   D   3,000

8   Hui buys a bedroom set for 5,415 dollars. He makes a down payment of 542 dollars. The rest must be paid off in 24 equal monthly payments. About how much will each payment be?

   F   125 dollars
   G   130 dollars
   H   200 dollars
   J   350 dollars

9 Which of these groups of numbers is in order from least to greatest?

A 1,052 5,300 5,067 5,302 15,090
B 1,052 5,067 5,300 5,302 15,090
C 5,300 5,067 5,302 1,052 15,090
D 5,300 5,067 5,302 15,090 1,052

10 What does the 8 in 1,892,035 represent?

F 8 thousands
G 8 ten-thousands
H 8 hundred-thousands
J 8 million

11 Mia counted 15 dandelions growing in just one typical square yard of her lawn. What else does she need to know before she can estimate the total number of dandelions in the entire lawn?

A the average size of each dandelion
B the number of seeds produced by each dandelion
C the total size of her lawn, in square yards
D the number of inches in a yard

12 Vic saves $50 to $200 each month. At this rate, how long will it take him to save $1,000?

F 20 to 50 months
G 5 to 20 months
H 2 to 5 months
J 5 to 10 months

13 If you are estimating by rounding to the nearest ten dollars, what numbers should you use to estimate $143 + $59 + $89?

A $143, $59, and $89
B $140, $60, and $90.00
C $143, $60, and $89.00
D $150, $60, and $90.00

14 What number is ten thousand more than 14,567,349?

F 14,567,449
G 14,568,349
H 14,577,349
J 14,667,349

15 Air Tite weather stripping comes in 75-foot rolls. Bianca needs 220 feet to go around her windows and another 50 feet to go around her doors. Which of these shows one way Bianca can figure out how many rolls of weather stripping she needs altogether?

A Add 220 feet and 50 feet. Then subtract 75 feet from the sum.
B Divide 220 feet by 75 feet. Then add 50 feet.
C Add 220 feet and 50 feet. Then divide the sum by 75.
D Add 220 feet, 75 feet, and 50 feet.

Decimal Place Values

There are **place values** to the right of the ones place. In a number such as 2.63, the period is called a **decimal point** and the first four places to the right of the decimal point are **tenths, hundredths, thousandths,** and **ten-thousandths.** The number 2.63 is called a decimal fraction or, more simply, a **decimal.**

In the decimal 2.63, the digit 6 represents six tenths and the digit 3 represents three hundredths.

When you say or write a decimal in words, you use the word **and** to represent the decimal point. So the decimal 2.63 is *"two **and** sixty-three hundredths."*

Amounts of money use decimals. A cent is a special name for *one hundredth of a dollar.*

> The 4 in $9.42 refers to four tenths of a dollar.
>
> The 2 in $9.42 is two hundredths of a dollar.

PRACTICE

Fill in the blanks below.

1 In the number 0.051, what place is the 1 in? _____

2 In the number 1.12596, what place is the 9 in? _____

3 In the number 0.00009, what place is the 9 in? _____

4 The number 0.01 is one ? . _____

5 The number 0.005 is five ? . _____

6 The number 0.003, written in words, is ? . _____

7 The number sixty-five hundredths, in digits, is ? . _____

8 The number four hundred three thousandths, in digits, is ? . _____

To round a decimal to a place value, look at the digit to the right of that place value.
   If the digit to the right is 5 or more, round up. If the digit to the right is 4 or less, round down.

Examples:
   0.0916 rounded to the tenths place is 0.1.
   0.0916 to the hundredths place is 0.09.
   0.0916 to the thousandths place is 0.092.

9 The number 0.6135, rounded to the nearest tenth, is ? . _____

10 The number 5.1058, rounded to the nearest thousandth, is ? . _____

Comparing Decimal Numbers

To compare two decimal numbers, start by lining up the decimal points. Then, moving from left to right, compare the digits.

$$0.0\ 2\ 7 \qquad\qquad 2.1\ 0\ 5 \qquad\qquad 1\ 5.0\ 6\ 3$$
$$\updownarrow \qquad\qquad\qquad \updownarrow \qquad\qquad\qquad\quad \updownarrow$$
$$0.0\ 0\ 8 \qquad\qquad 1\ 2.1\ 8\ 2 \qquad\qquad 1\ 5.0\ 1\ 7$$

2 is larger than 0.   1 is larger than 0.   6 is larger than 1.
The top number is larger.   The bottom number is larger. The top number is larger.

As another example, compare 0.1 and 0.008. You can think of 0.1 as 0.100. In the tenths place, 1 is larger than 0. That means 0.1 is larger than 0.008. With decimals, any nonzero digit *close to* a decimal point represents a larger value than any digit *farther to the right* of the decimal point.

PRACTICE

1 The number 1 is
0.01 times __?__ .   _____

2 The number 0.1 is
0.01 times __?__ .   _____

3 The number 0.01 is
0.001 times __?__ .   _____

4 Which number is larger, 1
or 0.01 times 80?   _____

5 Write 0.03 as a fraction.   _____

6 Write 0.15 as a fraction.   _____

7 Write 0.034 as a fraction.   _____

8 Write 0.0903 as a fraction.   _____

9 Write $\frac{37}{100}$ as a decimal.   _____

10 Write $\frac{9}{100}$ as a decimal.   _____

11 Write $\frac{15}{1000}$ as a decimal. _____

12 Write $\frac{3}{10000}$ as a decimal. _____

Circle the larger number in each set.

13 1   0.91

14 1   1.013

15 12.023   12.059

16 0.0607   0.0092

17 0.16   0.0082467

Circle the largest number in each set.

18 0.05   0.005   0.5

19 1.13   1.1   1.135

20 7.429   7.42   0.7429

Rewrite each set of numbers in order from *smallest to largest*.

21 0.6, 6, 0.06

22 0.018, 0.01723, 0.22, 10.28

23 0.891, 0.091, 0.081

Adding Decimals

To add the numbers **0.0013, 29,** and **1.51,** start by writing the numbers in column form with the decimal points lined up. For the whole number **29,** write a decimal point just to the right of the ones place.

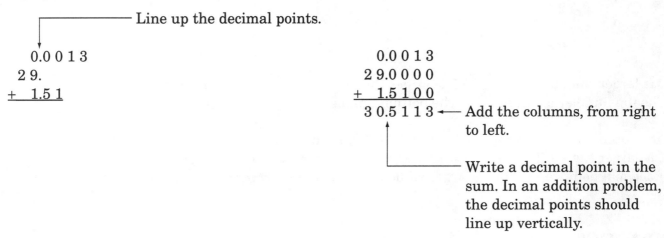

Write a zero in each empty column. Then add the columns from right to left. When you are finished adding each column, write a decimal point in your answer. All the decimal points should line up vertically.

PRACTICE

Add and then check your answers using rounded values. Be sure to start with the problem written in column form, and write a decimal point after each whole number.

1   46 m + 1.75 m =

_____

2   0.003 + 0.017 = _____

3   What is 5.302 million plus 0.0309 million?

4   0.0703 + 0.1506 =

_____

5   0.19 g + 0.015 g =

_____

6   0.018 + 0.75 + 1.25 =

_____

7 $0.09 + $0.13 + $0.53 =

_____

8 $17 + $2.05 + $0.04 =

_____

9   129 + 8.1 + 0.03011 =

_____

10   0.6 + 50 + 12.42 =

_____

11   1.324 + 0.085 = _____

12   0.59 + 0.01 = _____

Subtracting Decimals

To subtract decimals, start by writing the numbers in column form, with the decimal points lined up. Write a zero in each empty column, then subtract. The decimal point in the difference should line up with the other decimal points.

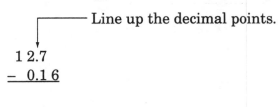

Line up the decimal points.

$$\begin{array}{r} 12.7 \\ -\ 0.16 \end{array}$$

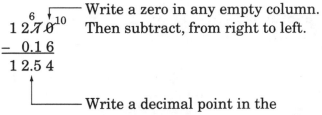

Write a zero in any empty column. Then subtract, from right to left.

$$\begin{array}{r} 12.70 \\ -\ 0.16 \\ \hline 12.54 \end{array}$$

Write a decimal point in the difference. In a subtraction problem, the decimal points should line up vertically.

To subtract from a row of zeros, a shortcut is to write a digit one less than the first nonzero digit from the right. Next, write a ten over the zero farthest to the right and write 9 over the other zeros. Then you are ready to subtract.

$$\begin{array}{r} 24.0000 \\ -\ 0.0028 \end{array} \qquad \begin{array}{r} 24.0000 \\ -\ 0.0028 \\ \hline 23.9972 \end{array}$$

PRACTICE

Subtract and then check your answers using addition.

1   14.091 − 0.050 = _____

2   0.157 − 0.03 = _____

3   21.4 − 0.025 = _____

4   0.145 − 0.073 = _____

5   0.9005 − 0.0814 = _____

6   5 − 0.057 = _____

7   What is 0.89 miles subtracted from 7 miles?

8   What is 0.0105 grams subtracted from 4.202 grams?

9   What is twenty-five dollars minus three cents?

10 What is 47 cents subtracted from $1.15?

11 What is 26.4 minus 0.0009?

12 What is 90 minus 0.008?

13 The Calgary Zoo's yearly budget is $14 million. The Granby Zoo in Quebec has a yearly budget of $4.5 million. How much more does the Calgary Zoo spend each year than the Granby Zoo?

14 A human being's top running speed is 27.89 miles per hour. A greyhound can run as fast as 39.35 miles per hour. How much faster is a greyhound than a human?

Multiplying Decimals

When you multiply decimals, at first you ignore the decimals points. Write the problem with the numbers lined up at the right. After you multiply, count the number of digits to the right of each decimal point. That is the **number of decimal places** for each number.

The number of decimal places in the product must equal the sum of the decimal places in the numbers you multiplied.

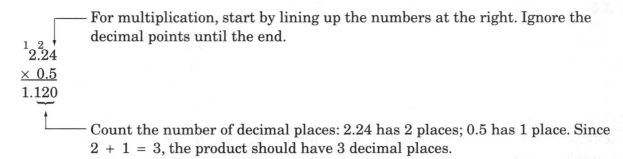

For multiplication, start by lining up the numbers at the right. Ignore the decimal points until the end.

Count the number of decimal places: 2.24 has 2 places; 0.5 has 1 place. Since 2 + 1 = 3, the product should have 3 decimal places.

In the problem below, 0.15 has two decimal places and 0.03 has two decimal place. That means the product must have 2 + 2 = 4 decimal places. To write the product, you have to write two zeros *to the left* of the nonzero digits.

$$
\begin{array}{r} 0.1\,5 \\ \times\ 0.0\,3 \\ \hline 4\,5 \end{array}
\qquad
\begin{array}{r} 0.1\,5 \\ \times\ \ 0.0\,3 \\ \hline 0.0\,0\,4\,5 \end{array}
$$

The product should have 4 decimal places. Write two zeros to the left of "45".

PRACTICE

For questions 1–12, rewrite the product for each multiplication problem by writing the answer with a decimal point.

1
$$
\begin{array}{r} 0.02 \\ \times\ 0.3 \\ \hline 6 \end{array}
$$

2
$$
\begin{array}{r} 0.4 \\ \times\ 0.009 \\ \hline 36 \end{array}
$$

3
$$
\begin{array}{r} 0.0052 \\ \times\ \ \ \ 2 \\ \hline 104 \end{array}
$$

4
$$
\begin{array}{r} 67 \\ \times\ 0.05 \\ \hline 335 \end{array}
$$

5
$$
\begin{array}{r} 50.12 \\ \times\ 0.4 \\ \hline 20048 \end{array}
$$

6
$$
\begin{array}{r} 11.2 \\ \times\ 0.4 \\ \hline 448 \end{array}
$$

7
$$
\begin{array}{r} 0.015 \\ \times\ 0.005 \\ \hline 75 \end{array}
$$

8
$$
\begin{array}{r} 0.05 \\ \times\ 0.05 \\ \hline 25 \end{array}
$$

9
$$
\begin{array}{r} 46,789 \\ \times\ 0.001 \\ \hline 46789 \end{array}
$$

10
$$
\begin{array}{r} 679.9 \\ \times\ 0.001 \\ \hline 6799 \end{array}
$$

11
$$
\begin{array}{r} 345 \\ \times\ 0.1 \\ \hline 345 \end{array}
$$

12
$$
\begin{array}{r} 5.156 \\ \times\ 0.1 \\ \hline 5156 \end{array}
$$

For questions 13 and 14, look for patterns in problems 9 through 12.

13 To multiply a number by 0.1, move the decimal point one place to the __?__ . _____

14 To multiply a number by 0.001, move the decimal point __?__ places to the __?__ . _____

Solve the problems below. Then look over your work to make sure it is correct. *Hint:* If there are any zeros at the ends of answers, write them down. You may erase them *after* you put in the decimal point.

15 0.03 × 0.4 = _____

16 6 × 0.21 = _____

17 10 × 5.3 = _____

18 4.5 × 0.3 = _____

"Of" *and* "×"
$\frac{1}{2}$ **of 10 means** $\frac{1}{2}$ **× 10**
0.4 of 20 means 0.4 × 20

---

19 What is five tenths of 0.011?

20 What is twelve hundredths of 56?

21 What is seventy-five thousandths of 20?

When you multiply a dollar amount by another decimal, the exact answer may have more than two decimal places. These answers must be rounded off to the nearest cent.

Multiply the dollar amounts below. Round your answer if necessary.

22 $11.09 × 3 = _____

23 $5.10 × 12 = _____

24 $104.00 × 0.22 = _____

25 $25.95 × 1.6 = _____

26 Find four tenths of $86.00.

27 Find fifteen hundredths of $42.00.

Set up and solve each word problem below.

28 Wayne's agent takes one tenth of the money from all contracts the agent supplies. When Wayne sings Saturday night, he will earn $560.00. How much money will the agent get?

29 Broccoli costs 89 cents a pound. You select a bunch that weighs 2.4 pounds. How much will it cost?

30 The sales tax in your area is 9 percent (or nine hundredths). How much tax must you pay on a $52.00 purchase?

31 Ricky is part of a 10-person lottery pool. The pool wins $2,750.00. What is one tenth of $2,750.00?

32 There are 27 million U.S. households with cats. On average, each of those houses has 2.2 cats. About how many pet cats are there in the U.S.?

Dividing a Decimal by a Whole Number

To divide a decimal by a whole number, start by writing a decimal point above the decimal point in the dividend. Then divide as usual.

Problem: 0.216 ÷ 3 = _____

$$\frac{.}{3\overline{)0.216}}$$	$$\frac{0.0}{3\overline{)0.216}}$$	$$\frac{0.072}{3\overline{)0.216}}$$
Write a decimal point directly above the decimal point inside the division bracket.	Write 0 above the 0 and above the 2, because 3 is greater than 2.	Write 7 above the 1 because 21 ÷ 3 = 7. Write 2 above the 6 because 6 ÷ 3 = 2.

Inside the division symbol, you can add zeros at the right of a decimal. That lets you keep dividing.

$$\frac{2.0}{6\overline{)12.3}}$$ Divide 6 into 12. The quotient digit is 2. Divide 6 into 3. The quotient digit is 0.	Write a zero to the right of the 3 to keep dividing.	$$\frac{2.05}{6\overline{)12.30}}$$ Divide 6 into 30. The quotient digit is 5.

PRACTICE

Find each quotient. Check your answers using multiplication.

1 $8\overline{)0.164}$

2 $5\overline{)15.25}$

3 $4\overline{)24.52}$

4 $3\overline{)\$15.24}$

5 $7\overline{)\$211.61}$

6 $6\overline{)3.372}$

7 $14\overline{)4.2}$

8 $21\overline{)5.25}$

9 $17\overline{)0.391}$

10   What is 0.093 ÷ 3?

11   What is 10.205 divided by 5?

12   What is $57.14 ÷ 30? (*Remember:* Round dollar amounts to the nearest cent.)

Dividing a Decimal by a Decimal

To divide a decimal by a decimal, start by moving both decimal points until the divisor is a whole number. Be sure to move both decimal points the *same* number of places in the *same* direction.

$$0.33\overline{)0.429} \qquad 33.\overline{)042.9} \qquad$$ Move both decimal points until the divisor is 33, a whole number.

$$33.\overline{)42.9} \qquad$$ Then write a decimal point for the quotient.

$$\begin{array}{r} 1.3 \\ 33.\overline{)42.9} \\ \underline{33} \\ 99 \\ \underline{99} \\ 0 \end{array}$$ Divide 33 into 42.9. The quotient is 1.3.

As another example, you know that $\dfrac{50}{0.08} = \dfrac{50 \times 100}{0.08 \times 100} = \dfrac{5000}{8}$.

So the division problem $0.08\overline{)50}$ has the same answer as $008.\overline{)5000}$ or $8\overline{)5000}$.

PRACTICE

Find each quotient. Use multiplication to check your work.

1 $1.5\overline{)66}$

2 $0.07\overline{)42}$

3 $0.25\overline{)2.8}$

4 $0.03\overline{)\$25.89}$

5 $0.06\overline{)762}$

6 $0.022\overline{)682}$

7 $0.91\overline{)18.2}$

8 $0.03\overline{)0.6351}$

9 $0.9\overline{)63}$

10 $0.33\overline{)693}$

11 You pay $6.15 for 1.5 pounds of steak. How much did you pay per pound?

12 Roses cost 89 cents each. How many can you buy for $17.80?

13 A 34.5-foot fence must be divided into 1.5-foot sections. How many sections will there be?

Adding Zeros at the Right of a Decimal

For the division problem 20 ÷ 8, one way to write the quotient is 2 r 4, as shown below at the left. However, if you write a zero after the decimal point in the dividend 20, as shown below at the right, you can see that another way to write the quotient is 2.5.

$$
\begin{array}{r}
2. \\
8\overline{)20.} \\
\underline{16} \\
4
\end{array}
\qquad\qquad
\begin{array}{r}
2.5 \\
8\overline{)20.0} \\
\underline{16} \\
40 \\
\underline{40} \\
0
\end{array}
$$

Sometimes, no matter how many zeros you add at the right of a decimal, the division process does not stop. If this happens, you usually should round the quotient to the nearest tenth, hundredth, or thousandth.

$$
\begin{array}{r}
1.48333 \\
60\overline{)89.00000} \\
\underline{60} \\
290 \\
\underline{240} \\
500 \\
\underline{480} \\
200 \\
\underline{180} \\
200 \\
\underline{180} \\
20
\end{array}
$$

PRACTICE

In each problem below, add zeros to the right of the decimal point of the dividend (the number inside the division symbol). Divide to the thousandths place. Then round your answer to the hundredths place.

Hint: $0.333\ldots = \frac{1}{3}$

$0.6666\ldots = \frac{2}{3}$

1 $4\overline{)6}$

2 $4\overline{)7}$

3 $3\overline{)35}$

4 $9\overline{)47}$

5 $11\overline{)39}$

6 $7\overline{)40}$

7 $11\overline{)45}$

8 $6\overline{)89}$

9 $12\overline{)4}$

Solving Mixed Word Problems

Set up and solve each word problem below. Some problems can be solved in one step. Some problems need two steps or even three steps to find a solution. *Remember:* All dollar amounts should be rounded to the nearest cent.

1   The average worker in Mexico earns $1.51 per hour. The wage of the average worker in the U.S. is 11.5 times higher. How much does the average U.S. worker earn per hour?

2   A farm produces about 3,000 eggs each day and sells them for $0.80 *per dozen.* On average, how much money does the farm get for its eggs each day?

3   Gorgio paid $5.33 for 1.3 pounds of beef. How much did he pay per pound?

4   Kyoko is making an iron support for her grape plants. She needs 2 long lengths of pipe each 3.25 feet long. She also needs a shorter length of pipe 2.5 feet long. How much pipe does she needs altogether.

5   Charlene runs a 3.4-mile course three times a week. How many miles does she run in a year? (There are 52 weeks in a year.)

6   When Dan began driving to Cleveland, the odometer on his car read exactly 2,342.8 miles. When he got to Cleveland, it read 2,690.2. How many miles was the trip?

7   You rent a car from a company that charges you 43 cents per mile. When you pick up the car, the odometer reads 11,542.5. When you return the car, the odometer reads 11,562.7. How much will you be charged for mileage?

8   You are going on a tour of a local microbrewery in a group of 22 people. The tour costs $2.50 per person, and it will cost $24.00 to rent the bus. If you split all costs evenly among the 22 people, how much will each person pay?

9   Larry bought a large rug for $650.00. He made a down payment of $98.00. The remainder was paid off in 12 equal payments. How much was each payment?

Decimals Skills Practice

Circle the letter for the correct answer to each problem. Try crossing out unreasonable answers before you start to work.

1   0.851 + 0.08 = _____
   - **A**   0.931
   - **B**   1.651
   - **C**   0.859
   - **D**   9.31
   - **E**   None of these

2

11.57 − 0.7 = _____
   - **F**   10.67
   - **G**   11.87
   - **H**   10.87
   - **J**   11.50
   - **K**   None of these

3

25.02 × 30 = _____
   - **A**   75.6
   - **B**   75,060
   - **C**   75.06
   - **D**   750.6
   - **E**   None of these

4

$3.1\overline{)992}$
   - **F**   320
   - **G**   32
   - **H**   3.2
   - **J**   33
   - **K**   None of these

5

2.5 ÷ 0.05 = _____
   - **A**   5
   - **B**   0.5
   - **C**   0.05
   - **D**   50
   - **E**   None of these

6   Which group of numbers is in order from least to greatest?
   - **F**   5.619, 5.61, 5.19, 5.2
   - **G**   5.61, 5.619, 5.19, 5.2
   - **H**   5.19, 5.61, 5.619, 5.2
   - **J**   5.19, 5.2, 5.61, 5.619

7   You are estimating by rounding to the nearest whole number. What numbers should you use to divide 5.167 by 4.789?
   - **A**   5 and 5
   - **B**   5 and 4
   - **C**   5.2 and 4.8
   - **D**   5.1 and 4.7

8   There are 3.96 million people living in Cleveland and there are 2.2 million households. Which of these is the best estimate of the average number of people who live in each household?
   - **F**   2
   - **G**   1.5
   - **H**   1.3
   - **J**   1

9   It takes Ray Lynn about 12 minutes to make one pot holder, and she sells each one for $5.50. How much does she make per hour? (There are 60 minutes in one hour. Ignore her costs.)
   - **A** $6.60
   - **B** $25.83
   - **C** $27.50
   - **D** $11.00

10   One box of cereal weighs 0.781 pounds. Each carton contains 18 boxes of cereal. Which of these is the best estimate of how much each carton weighs?
   - **F**   10.5 pounds
   - **G**   14 pounds
   - **H**   7.8 pounds
   - **J**   18 pounds

11

$109.56
+ 98.90

A $208.46
B $198.46
C $207.46
D $308.46
E   None of these

12

99.081
−87.809

F   12.272
G   0.11272
H   11.282
J   11.272
K   None of these

13 $21.52 × 0.5 = _____

A $107.60
B $10.76
C $10.75
D $17.60
E   None of these

14

1.872 ÷ 12 = _____

F   1.06
G   106
H   0.17 r 2
J   156
K   None of these

15

5)192

A   39
B   38.4
C   37.4
D   3.84
E   None of these

16   Which decimals can be added to 0.07 to make a sum that is greater than 1?

F   any decimal greater than 0.3
G   any decimal greater than 0.03
H   any decimal greater than 0.33
J   any decimal greater than 0.93

17   The new cash registers at Bob's Burger Barn found that the clerks spent an average of 3.456908 minutes with each customer. Bob reported this number as 3.5 minutes. To what place did he round the number?

A   ones
B   thousandths
C   tenths
D   fifths

18   What does the 3 in 1.80367 represent?

F   3 hundredths
G   3 thousands
H   3 tenths
J   3 thousandths

19   Which of these number sentences is true?

A   0.013 < 0.0109
B   0.782 < 0.882
C   1 < 0.789
D   0.45 < 0.097

Fractions

Numerators and Denominators

In a fraction, the number on the bottom tells how many parts the object or amount has been divided into. The bottom number is called the **denominator.** The top number in a fraction is called the **numerator.** It refers to the number of parts that are shown or shaded.

Darlene has 59 pages to type and so far she has typed 23 pages. She has finished $\frac{23}{59}$ of the job.

There are 500 people expected at Tuesday's concert and there are 1,000 people expected at Friday's concert. There are $\frac{1}{2}$ as many people expected at Tuesday's concert as at Friday's concert.

PRACTICE

Use the Civil War statistics to answer questions 1 through 4. You will have to add and subtract to find some of the numbers for these fractions. *Remember:* The total for each comparison goes in the denominator.

Civil War Statistics	
Number of men who served:	2,213,582
Casualties:	646,392
died of war wounds:	140,415
died of other causes:	224,097
wounded, but survived	281,881

1   What fraction of Civil War soldiers became casualties? _____

2   What fraction of Civil War soldiers died during the war? _____

3   The soldiers who died from battle wounds during the Civil War were what fraction of the number who died from other causes? _____

4   The number of soldiers who died or were wounded during the Civil War was what fraction of the number who survived unharmed? _____

Use this information to answer questions 5 through 8. For now, do not simplify any fractions.

Last year, Tomas's business made a little over $250,000. Thomas paid about $107,800 in taxes. He paid his employees a total of about $63,000. His other expenses were approximately $45,000.

5   What fraction of the income of the business went to pay taxes? _____

6   What fraction of the income was used for employee's wages and other expenses? _____

7   The amount Tomas spends on "other expenses" is what fraction of the amount he spends to pay employees? _____

8   What fraction of the income of the business did Tomas keep? _____

Finding a Fraction of a Number

It is important to understand the relationship between fractions, multiplication, and division. **For example:**

$$\frac{1}{3} \text{ of } 312 = \frac{1}{3} \times 312 = \frac{312}{3} \text{ or } 3\overline{)312}.$$

$$\frac{1}{6} \text{ of } 600 = \frac{1}{6} \times 600 = \frac{600}{6} \text{ or } 6\overline{)600}.$$

Also, $\frac{11}{20}$ can be written as $11 \div 20$ *or* $20\overline{)11}$.

This suggests one way to write a fraction as a decimal: Divide the numerator in the fraction by the denominator.

$$\frac{1}{4} = 4\overline{)\begin{array}{r} 0.25 \\ 1.00 \\ \underline{8} \\ 20 \\ \underline{20} \\ 0 \end{array}}$$

PRACTICE

Find each value below.

1 $\frac{1}{4}$ of 300

2 $\frac{2}{4}$ of 300

3 $\frac{1}{5}$ of 245

4 $\frac{3}{4}$ of 420

5 $\frac{5}{6}$ of 126

6 $\frac{7}{12}$ of 180

7 $\frac{5}{8}$ of 44

Solve each problem below.

8 One-fourth of the 5,248 students at Fairfield Technical School are women. How many women go to the school?

9 Four-fifths of the money spent in the state lottery goes to the public schools. If you spend $55.00 on lottery tickets, how much of your money goes to schools?

10 Leanne weighs $\frac{2}{3}$ as much as her sister. If Leanne's sister weighs 180 pounds, how much does Leanne weigh?

Write each fraction below as a decimal. Use a series of dots to show that a digit repeats. For example, the decimal "0.33..." represents a number where the digit 3 repeats over and over.

11 $\frac{1}{2}$

12 $\frac{1}{4}$

13 $\frac{1}{5}$

14 $\frac{3}{4}$

15 $\frac{4}{12}$

16 $\frac{2}{3}$

Reducing a Fraction to Simplest Terms

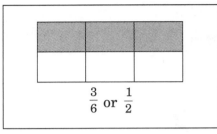

$$\frac{3}{6} \text{ or } \frac{1}{2}$$

There are several ways to name the fraction that represents this figure. If you think of the figure as being divided into six parts, then $\frac{3}{6}$ is shaded. If you think of it as being divided into two parts, then $\frac{1}{2}$ is shaded. The fractions $\frac{3}{6}$ and $\frac{1}{2}$ have the same value. Then are called are **equivalent fractions.**

When you reduce a fraction, you change it into an equivalent fraction with smaller numbers. It is easier to understand and use fractions when they have been reduced. **To reduce a fraction, divide both the top and the bottom by the same number.**

The fraction $\frac{3}{9}$ can be reduced further by dividing the numerator 3 and the denominator 9 by 3:

$$\frac{3}{9} = \frac{3 \div 3}{9 \div 3} = \frac{1}{3}$$

Problem: Reduce $\frac{12}{30}$.

Both the numerator 12 and the denominator 30 can be evenly divided by 2: $\qquad \frac{12}{30} = \frac{12 \div 2}{30 \div 2} = \frac{6}{15}$

The new numerator 6 and denominator 15 can be evenly divided by 3: $\qquad \frac{6}{15} = \frac{6 \div 3}{15 \div 3} = \frac{2}{5}$

No number can evenly divide both 2 and 5, so the fraction $\frac{2}{5}$ cannot be reduced any more. It is in **lowest terms** or **simplest terms.** This is also called **reducing a fraction.**

PRACTICE

Reduce each fraction to simplest terms.

Hint: any proper fraction with a prime number in the denominator is in simplest terms. The first nine prime numbers are 1, 2, 3, 5, 7, 11, 13, 17, 19. So fractions such as $\frac{1}{5}$, $\frac{3}{5}$, and so on are in simplest form.

1	$\frac{6}{12}$	6	$\frac{75}{125}$	11	$\frac{15}{90}$	16	$\frac{22}{110}$
2	$\frac{15}{45}$	7	$\frac{132}{420}$	12	$\frac{22}{64}$	17	$\frac{30}{150}$
3	$\frac{9}{27}$	8	$\frac{65}{145}$	13	$\frac{25}{200}$	18	$\frac{20}{800}$
4	$\frac{70}{100}$	9	$\frac{5}{250}$	14	$\frac{24}{60}$	19	$\frac{25}{500}$
5	$\frac{24}{72}$	10	$\frac{33}{90}$	15	$\frac{180}{360}$		

Reducing a Fraction to Simplest Terms

Fractions Equal To 1 and Fractions Greater Than 1

So far, you have worked with fractions such as $\frac{4}{5}$ or $\frac{1}{3}$, where the top number is smaller than the bottom number. These fractions are called **proper fractions.** If the top number in a fraction is equal to or larger than the bottom number, as in examples such as $\frac{5}{4}$, $\frac{7}{6}$, and $\frac{9}{9}$, the fraction is called an **improper fraction.**

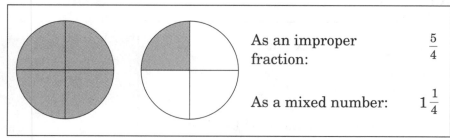

In an improper fraction, the numerator is equal to or greater than the denominator. An improper fraction represents 1 unit or a number larger than 1 unit.

As an improper fraction: $\frac{5}{4}$

As a mixed number: $1\frac{1}{4}$

Here are some other numbers that are greater than one: $3\frac{3}{4}$, $6\frac{1}{2}$, $1\frac{1}{4}$, and so on. Each of these is a **mixed number.** A mixed number is the sum of a whole number and a proper fraction.

To change an improper fraction to a whole number, start by dividing the top number by the bottom number. If there is no remainder, your answer is a whole number.

$$\frac{8}{2} = 8 \div 2 = 4$$

If there is a remainder, then your answer is a mixed number. For the fraction, the numerator is the remainder and the denominator is the denominator of the original improper fraction.

$$\frac{7}{5} = 7 \div 5 = 1 \text{ r } 2$$

$$\text{So } \frac{7}{5} = 1\frac{2}{5}.$$

PRACTICE

In each box below, write a ">" sign, a "<" sign, or an "=" sign to compare each fraction to the whole number 1.

1. $\frac{3}{5}$ ☐ 1

2. $\frac{6}{3}$ ☐ 1

3. $\frac{9}{7}$ ☐ 1

4. $\frac{12}{12}$ ☐ 1

5. $\frac{3}{16}$ ☐ 1

6. $1\frac{2}{5}$ ☐ 1

Write each improper fraction as a whole number or as a mixed number. Reduce all proper fractions to simplest terms.

7. $\frac{125}{25} = $ ___

8. $\frac{5}{5} = $ ___

9. $\frac{6}{4} = $ ___

10. $\frac{12}{3} = $ ___

11. $\frac{10}{7} = $ ___

12. $\frac{7}{3} = $ ___

13. $\frac{9}{3} = $ ___

14. $\frac{15}{4} = $ ___

15. $\frac{16}{8} = $ ___

16. $\frac{17}{8} = $ ___

17. $\frac{55}{20} = $ ___

18. $\frac{43}{43} = $ ___

19. $\frac{75}{50} = $ ___

20. $\frac{67}{67} = $ ___

21. $\frac{50}{10} = $ ___

Adding and Subtracting Like Fractions

Two fractions are "like" if they have the same denominator. They are "unlike" if they have different denominators. To add or subtract two fractions that have the same denominator, use that same denominator and add or subtract the two numerators. Reduce the result if it is not in lowest terms. *Remember:* Zero divided by any other number equals zero.

Add:

$$\frac{3}{8} + \frac{1}{8} = \frac{3+1}{8} \longleftarrow \text{Add the two numerators.}$$

$$= \frac{4}{8}$$

$$= \frac{4}{4} \times \frac{1}{2} \quad \text{Reduce the fraction.}$$

$$= \frac{1}{2}$$

Subtract:

$$\frac{9}{15} - \frac{4}{15} = \frac{9-4}{15} \longleftarrow \text{Subtract the two numerators.}$$

$$= \frac{5}{15}$$

$$= \frac{5 \div 5}{15 \div 5} \quad \text{Reduce the fraction.}$$

$$= \frac{1}{3}$$

To add two mixed numbers, combine the fractions and simplify the sum. Then add the whole numbers.

$$
\begin{array}{r}
{}^1 3\frac{3}{4} \\
+\, 2\frac{3}{4} \\
\hline
6\frac{1}{2}
\end{array}
$$

Add the fractions:

$$\frac{3}{4} + \frac{3}{4} = \frac{6}{4} = 1\frac{1}{2}$$

Write $\frac{1}{2}$. Carry the 1.

$$1 + 3 + 2 = 6$$

PRACTICE

Add or subtract as indicated, simplifying all answers. Subtract to check your work on the addition problems. Add to check the subtraction problems.

1 $\frac{3}{4} + \frac{1}{4} =$ _____

2 $\frac{4}{5} + \frac{2}{5} =$ _____

3 $\frac{5}{8}$ cup $- \frac{4}{8}$ cup = _____

4 $\frac{7}{12} - \frac{5}{12} =$ _____

5 $4\frac{3}{10}$ m $+ 1\frac{2}{10}$ m = ___

6 $\frac{12}{15} - \frac{2}{15} =$ _____

7 $1\frac{3}{4} - 1\frac{1}{4} =$ _____

8 $2\frac{1}{3} + 3\frac{1}{3} =$ _____

9 $\frac{3}{6} + \frac{1}{6} + \frac{4}{6} =$ _____

10 $\frac{9}{16}$ mi $- \frac{5}{16}$ mi = _____

11 $\frac{10}{21} - \frac{2}{21} - \frac{3}{21} =$ _____

12 $\frac{4}{5} + \frac{3}{5} + \frac{1}{5} =$ _____

13 $\frac{2}{3}$ ft $+ \frac{2}{3}$ ft = _____

14 A recipe calls for $\frac{1}{3}$ cup water and $\frac{2}{3}$ cup milk. How much liquid does it contain?

15 Sheryl has $\frac{3}{4}$ of a cup of milk. If she used $\frac{1}{4}$ cup, how much will she have left?

To add two mixed numbers, add the fractions and simplify the sum. Then add the whole numbers.

If the sum of the fractions is a mixed number, carry the whole number to the column of whole numbers.

$$^{1}3\frac{3}{4}$$
$$+2\frac{3}{4}$$
$$\overline{6\frac{1}{2}}$$

Add the fractions:
$$\frac{3}{4}+\frac{3}{4}=\frac{6}{4}=1\frac{1}{2}$$

Write $\frac{1}{2}$. Carry the 1.

$1+3+2=6$

To subtract mixed numbers, start by subtracting the fractions. You may need to borrow 1 from the whole number. Then subtract the whole numbers.

$$^{7}\cancel{8}\frac{11}{8}$$
$$-2\frac{7}{8}$$
$$\overline{5\frac{1}{2}}$$

$$8\frac{3}{8}=7+\frac{8}{8}+\frac{3}{8}=7\frac{11}{8}$$

Subtract the fractions.
$$\frac{11}{8}-\frac{7}{8}=\frac{4}{8}=\frac{1}{2}$$

Subtract the whole numbers.
$$7-2=5$$

PRACTICE

Carry or borrow to solve each problem below. Use addition to check the subtraction problems. Use subtraction to check the addition problems.

16 $3\frac{1}{4}-\frac{3}{4}=$ _____

17 $1\frac{3}{5}+\frac{3}{5}=$ _____

18 $1\frac{5}{9}-\frac{7}{9}=$ _____

19 $3\frac{7}{8}+2\frac{1}{8}=$ _____

20 $9\frac{1}{3}-7\frac{2}{3}=$ _____

21 $3\frac{5}{12}-\frac{9}{12}=$ _____

22 $5\frac{4}{5}+\frac{2}{5}=$ _____

23 $2\frac{3}{7}+2\frac{5}{7}=$ _____

24 $3-\frac{7}{10}=$ _____

25 $3\frac{4}{15}-\frac{7}{15}=$ _____

26 $5\frac{3}{8}+5\frac{7}{8}=$ _____

27 $10-9\frac{3}{4}=$ _____

28 Kayla spent $1\frac{1}{4}$ hour making dinner and $\frac{3}{4}$ hour cleaning up. How much time did she spend altogether working on the meal?

29 The Nature Valley Trail is $\frac{3}{5}$ of a mile long. The Covered Bridge Trail is $\frac{4}{5}$ of a mile long. If you hike both trails, how far will you walk?

Adding and Subtracting Like Fractions

Adding and Subtracting Unlike Fractions

Two fractions are "like" if they have the same denominator. They are "unlike" if they have different denominators. To add or subtract unlike fractions, start by rewriting them with a common denominator.

Example 1:

Rewrite $\frac{5}{6}$ and $\frac{11}{12}$ as like fractions.

Solution:

A common denominator for $\frac{5}{6}$ and $\frac{11}{12}$ is twelfths, because 12 is a **common multiple** of 6 and 12.

$$\frac{5}{6} = \frac{5 \times 2}{6 \times 2} = \frac{10}{12} \qquad \frac{11}{12} = \frac{11}{12}$$

The two fractions are $\frac{10}{12}$ and $\frac{11}{12}$.

Example 2:

Rewrite $\frac{2}{3}$ and $\frac{5}{8}$ as like fractions.

Solution:

A common denominator for $\frac{2}{3}$ and $\frac{5}{8}$ is twenty-fourths, because 24 is a **common multiple** of 3 and 8.

$$\frac{2}{3} = \frac{2 \times 8}{3 \times 8} = \frac{16}{24} \qquad \frac{5}{8} = \frac{5 \times 3}{8 \times 3} = \frac{15}{24}$$

The two fractions are $\frac{16}{24}$ and $\frac{15}{24}$.

Writing the fraction $\frac{5}{6}$ as $\frac{10}{12}$ is called **raising the fraction.** The two fractions $\frac{5}{6}$ and $\frac{10}{12}$ are called **equivalent fractions.**

PRACTICE

Write a number for each box so the two fractions are equivalent.

1 $\frac{2}{5} = \frac{\Box}{10}$

2 $\frac{1}{4} = \frac{\Box}{12}$

3 $\frac{1}{6} = \frac{\Box}{30}$

4 $\frac{1}{7} = \frac{\Box}{21}$

5 $\frac{2}{7} = \frac{\Box}{21}$

6 $\frac{3}{5} = \frac{\Box}{20}$

7 $\frac{5}{9} = \frac{\Box}{18}$

8 $\frac{1}{3} = \frac{\Box}{30}$

9 $\frac{1}{2} = \frac{\Box}{50}$

10 $\frac{1}{2} = \frac{\Box}{30}$

11 $\frac{5}{6} = \frac{\Box}{30}$

12 $\frac{1}{8} = \frac{\Box}{24}$

13 $\frac{8}{9} = \frac{\Box}{45}$

14 $\frac{3}{5} = \frac{\Box}{25}$

15 $\frac{5}{9} = \frac{\Box}{36}$

16 $\frac{3}{4} = \frac{\Box}{16}$

17 $\frac{3}{4} = \frac{\Box}{40}$

18 $\frac{5}{6} = \frac{\Box}{36}$

Finding the denominator 24 for the fractions $\frac{2}{3}$ and $\frac{5}{8}$ is called **finding the common denominator** of the fractions.

The **least common denominator** for two fractions is the smallest number that can be used as a denominator for the two fractions. For example, the two fractions $\frac{1}{2}$ and $\frac{3}{5}$ have common denominators of 10, 20, 30, 40, and so on. The denominator 10 is the smallest of the common denominators. As another example, the fractions $\frac{1}{3}$ and $\frac{4}{9}$ have common denominators of 9, 18, 27, and so on, but 9 is the least common denominator.

Find the least common denominator for each pair of fractions.

19 $\frac{1}{4}$ and $\frac{1}{2}$ _____

20 $\frac{1}{8}$ and $\frac{1}{4}$ _____

21 $\frac{1}{7}$ and $\frac{1}{3}$ _____

22 $\frac{1}{5}$ and $\frac{1}{3}$ _____

23 $\frac{3}{8}$ and $\frac{5}{6}$ _____

24 $\frac{1}{4}$ and $\frac{1}{5}$ _____

In each problem below, rewrite the fractions so they are *like fractions*. Then add or subtract as indicated. Be sure to write your answers in simplest terms.

25 $\begin{array}{r} \frac{1}{5} \\ + \frac{1}{4} \\ \hline \end{array}$

26 $\begin{array}{r} \frac{3}{7} \\ - \frac{1}{3} \\ \hline \end{array}$

27 $\begin{array}{r} \frac{2}{5} \\ + \frac{1}{10} \\ \hline \end{array}$

28 $\frac{5}{12} - \frac{1}{4} = $ _____

29 $\begin{array}{r} \frac{1}{3} \\ + \frac{5}{9} \\ \hline \end{array}$

30 $\begin{array}{r} \frac{7}{8} \\ + \frac{1}{2} \\ \hline \end{array}$

31 $\begin{array}{r} \frac{2}{3} \\ - \frac{4}{15} \\ \hline \end{array}$

32 $\frac{1}{2} - \frac{1}{10} = $ _____

33 $\frac{5}{6} + \frac{1}{3} = $ _____

34 $\frac{5}{9} - \frac{1}{4} = $ _____

35 $\frac{3}{5} - \frac{1}{3} = $ _____

36 Suli squeezes oranges until she has $\frac{5}{8}$ cup of orange juice. If he uses $\frac{1}{4}$ cup of the juice in a bread recipe, how much will be left?

Adding and Subtracting Unlike Fractions

Multiplying Fractions

To multiply two fractions, multiply the numerators together. Then multiply the denominators together.

$$\frac{2}{5} \times \frac{3}{4} = \frac{2 \times 3}{5 \times 4} = \frac{6}{20}$$ Multiply the numerators and multiply the denominators.

$$\frac{6}{20} = \frac{6 \div 2}{20 \div 2} = \frac{3}{10}$$ Reduce the fraction.

To multiply a fraction by a whole number, rewrite the whole number as an improper fraction with a denominator of 1. Then multiply the fractions.

$$\frac{2}{3} \times 6 = \frac{2}{3} \times \frac{6}{1}$$ Rewrite 6 as $\frac{6}{1}$.

$$= \frac{2 \times 6}{3 \times 1}$$ Multiply the fractions.

$$= \frac{12}{3}$$ Simplify the fraction.

$$= 4$$

PRACTICE

Multiply the fractions below. Reduce your answers to simplest terms.

1 $\frac{1}{3} \times \frac{4}{5} =$ _____

2 $\frac{1}{2} \times \frac{5}{8} =$ _____

3 $\frac{2}{5} \times \frac{1}{4} =$ _____

4 $\frac{5}{6} \times 10 =$ _____

5 $12 \times \frac{2}{3} =$ _____

6 $\begin{array}{r} \frac{3}{4} \\ \times \frac{2}{3} \\ \hline \end{array}$

7 $\begin{array}{r} \frac{1}{5} \\ \times 25 \\ \hline \end{array}$

8 $\begin{array}{r} \frac{5}{12} \\ \times 12 \\ \hline \end{array}$

9 $\begin{array}{r} \frac{3}{8} \\ \times \frac{1}{5} \\ \hline \end{array}$

10 $\frac{1}{3} \times \frac{1}{4} \times \frac{2}{3} =$ _____

Hint:
$$\frac{1}{3} \times \frac{1}{4} \times \frac{2}{3}$$
$$= \left(\frac{1}{3} \times \frac{1}{4} \right) \times \frac{2}{3}$$

11 $\frac{3}{4} \times \frac{2}{3} \times \frac{1}{2} =$ _____

12 $\frac{1}{5} \times \frac{4}{5} \times \frac{1}{3} =$ _____

13 What is $\frac{1}{3}$ of $\frac{8}{11}$?

14 What is $\frac{1}{2}$ of $\frac{3}{4}$ of a pound?

15 Leeland helps out at the shelter three days every week, even during holidays. If a "working week" is five working days, how many "working weeks" is Leeland at the shelter?

16 At Anne's office, $\frac{2}{3}$ of the people commute and $\frac{4}{5}$ of the commuters come from the city. What fraction of the people in Anne's office commute from the city?

To multiply two mixed numbers, write each of them as an improper fraction. Then multiply the two fractions.

Problem 1:

Write $3\frac{1}{8}$ as an improper fraction.

Solution:

$$3\frac{1}{8} = 3 + \frac{1}{8}$$
$$= \frac{24}{8} + \frac{1}{8}$$
$$= \frac{24 + 1}{8}$$
$$= \frac{25}{8}$$

Here is a shortcut. Multiply the whole number times the denominator, and then add the numerator. Write that result over the original denominator.

$$3\frac{1}{8} = \frac{8 \times 3 + 1}{8}$$
$$= \frac{25}{8}$$

Problem 2:

Find the product $2\frac{1}{5} \times 1\frac{2}{3}$.

Solution:

$$2\frac{1}{5} = \frac{10}{5} + \frac{1}{5} = \frac{11}{5}$$

$$1\frac{2}{3} = \frac{3}{3} + \frac{2}{3} = \frac{5}{3}$$

$$2\frac{1}{5} \times 1\frac{2}{3} = \frac{11}{5} \times \frac{5}{3}$$
$$= \frac{55}{15} = \frac{55 \div 5}{15 \div 5} = \frac{11}{3}$$

As a mixed number, $\frac{11}{3} = \frac{9}{3} + \frac{2}{3}$
$$= 3\frac{2}{3}$$

PRACTICE

Write each mixed number as an improper fraction.

17 $2\frac{1}{8} = $ _____

18 $3\frac{1}{4} = $ _____

19 $4\frac{1}{2} = $ _____

20 $3\frac{3}{8} = $ _____

21 $5\frac{1}{2} = $ _____

22 $12\frac{1}{4} = $ _____

23 $3\frac{7}{30} = $ _____

24 $5\frac{1}{5} = $ _____

25 $6\frac{4}{7} = $ _____

26 $7\frac{3}{5} = $ _____

27 $1\frac{4}{5} = $ _____

28 $1\frac{3}{4} = $ _____

Solve each multiplication problem.

29 $1\frac{1}{4} \times \frac{1}{5} = $ _____

30 $3\frac{2}{3} \times 1\frac{1}{2} = $ _____

31 $2\frac{1}{3} \times 2\frac{1}{2} = $ _____

32 $2\frac{1}{3} \times 1\frac{1}{5} = $ _____

Canceling Before You Multiply

To multiply $\frac{2}{5}$ times $\frac{3}{4}$, notice that the factors in the numerator and denominator have a common factor 2:

$$\frac{2}{5} \times \frac{3}{4} = \frac{2 \times 3}{5 \times 4} = \text{?}$$

A shortcut is to reduce the fractions before you multiply. This shortcut is called **canceling**.

To cancel, look for a common factor in the numerator and the denominator. Divide both numbers by the common factor. Then multiply as usual.

$$\frac{2}{5} \times \frac{3}{4} = \frac{\overset{1}{\cancel{2}} \times 3}{5 \times \cancel{4}_2}$$ Divide 2 into the numerator and divide 2 into the denominator.

$$= \frac{3}{10}$$ Then multiply the numerators and multiply the denominators.

In this problem, you can cancel two times.

$$\frac{3}{4} \times \frac{2}{3} \times \frac{1}{2} = \frac{\overset{1}{\cancel{3}} \times \overset{1}{\cancel{2}} \times 1}{4 \times \underset{1}{\cancel{3}} \times \underset{1}{\cancel{2}}}$$ Cancel by the factors 2 and 3.

$$= \frac{1}{4}$$

PRACTICE

Simplify the problems below by canceling. Then finish the multiplication or solve the problem.

1 $\frac{2}{15} \times \frac{3}{6} = $ _____

2 $\frac{3}{8} \times \frac{2}{3} = $ _____

3 $\frac{5}{12} \times \frac{4}{15} = $ _____

4 $\frac{1}{6} \times \frac{3}{4} \times \frac{2}{9} = $ _____

5 $\frac{1}{4} \times \frac{2}{3} \times \frac{3}{5} = $ _____

6 Damia's kitchen floor is 240 square feet in size. If she puts a wooden surface over $\frac{2}{5}$ of the floor, how much wooden flooring will she need?

7 At the Potoce Mine, visitors get to keep $\frac{1}{2}$ of any gold they find. Boris finds $\frac{4}{5}$ of an ounce of gold. How much does he get to keep?

8 One can of beans contains $1\frac{1}{2}$ cups. If three people share the beans, each get $\frac{1}{3}$ of the can. What is $\frac{1}{3}$ of $1\frac{1}{2}$ cups?

9 There are 120 members of an Athletic Club. In a poll, $\frac{5}{12}$ of the members said they swim, and $\frac{1}{5}$ of the swimers use the side stroke. How many club members use the side stroke?

Canceling Before You Multiply

Dividing with Fractions

The numbers 4 and $\frac{1}{4}$ are called **reciprocals** of each other. To find the reciprocal of a fraction, turn the fraction "upside down."

You know that $12 \div 4$ is the same as $12 \times \frac{1}{4}$. A rule is that dividing by a number is the same as multiplying by the reciprocal of that number. You can use the same rule for fractions: Dividing by a fraction is the same as multiplying by the reciprocal of that fraction.

If a division problem has a fraction and a whole number, start by rewriting the whole number as a fraction with a denominator of 1.

Example 1:

$$4 \div \frac{7}{8} = \frac{4}{1} \div \frac{7}{8}$$ Rewrite 4 as $\frac{4}{1}$.

$$= \frac{4}{1} \times \frac{8}{7}$$ Dividing by $\frac{7}{8}$ is the same as multiplying

$$= \frac{4 \times 8}{1 \times 7}$$ by $\frac{8}{7}$.

$$= \frac{32}{7}$$

$$= \frac{28}{7} + \frac{4}{7}$$ Multiply and simplify.

$$= 4\frac{4}{7}$$

Example 2:

$$\frac{2}{3} \div 5 = \frac{2}{3} \div \frac{5}{1}$$ Rewrite 5 as $\frac{5}{1}$.

$$= \frac{3}{2} \times \frac{1}{5}$$ Dividing by $\frac{5}{1}$ is the same as multiplying

  by $\frac{1}{5}$.

$$= \frac{3 \times 1}{2 \times 5}$$ Multiply and simplify.

$$= \frac{3}{10}$$

PRACTICE

Solve each division problem below. Reduce each answer to simplest terms.

1. $\frac{2}{3} \div \frac{1}{2} = $ _____

2. $\frac{1}{5} \div \frac{3}{7} = $ _____

3. $1\frac{1}{2} \div \frac{3}{8} = $ _____

4. $\frac{5}{8} \div 7 = $ _____

5. $\frac{6}{7} \div \frac{1}{5} = $ _____

6. $9 \div \frac{2}{3} = $ _____

7. $2\frac{1}{3} \div \frac{1}{3} = $ _____

8. $3\frac{3}{4} \div \frac{3}{4} = $ _____

9.   Trish bought $4\frac{1}{2}$ quarts of juice. She is going to put it into cups that hold $\frac{1}{2}$ cup (or $\frac{1}{8}$ quart) each. How many cups will she fill?

10.   Chuck has $\frac{3}{4}$ ton of gravel. He wants to divide it into $\frac{1}{8}$-ton piles. How many piles should he make?

11.   Alex works 7 hours a day at a barber shop. It takes him $\frac{3}{4}$ hour to cut one head of hair. How many haircuts can he give in one full workday? (*Hint:* Your answer should be the whole number of haircuts he can *complete*.)

Solving Mixed Word Problems

Solve each word problem below. Reduce all answers to simplest terms.

1 The Women's Club has sold 180 of the 360 circus tickets they need to sell. What fraction has been sold so far?

2 Each volume of an encyclopedia is $2\frac{1}{3}$ inches wide. How many volumes will fit on a shelf 18 inches wide?

3 Alana's doctor told her she should drink 7 glasses of water a day, each at least $1\frac{3}{4}$ cups in size. How many cups of water should Alana drink each day?

4 When the Patel family went apple picking, Mr. Patel picked $1\frac{1}{2}$ bushels, Mrs. Patel picked $\frac{3}{4}$ bushel, and each of the three children picked $\frac{1}{4}$ bushel. How many bushels of apples did the family pick?

5 One-fifth of the 110 members of Park Street Church are senior citizens. How many senior citizens belong to the church?

6 It usually takes Derek $2\frac{3}{4}$ hours to drive to Toronto. Today he got caught in a storm and it took $5\frac{1}{2}$ hours. How much time did the storm add to Derek's trip?

7 A sweater marked $65.20 is on sale for $\frac{1}{4}$ off. How much does it cost on sale? (*Hint:* This is a 2-step problem.)

8 Troy has one large can of chicken stock containing $26\frac{1}{2}$ ounces and three small cans of chicken stock containing $12\frac{1}{2}$ ounces each. How much chicken stock does he have altogether.

9 Tiffany uses $\frac{1}{10}$ of her wages each month to buy company stock. Also, she puts $\frac{1}{20}$ in a savings account and $\frac{1}{15}$ into the company pension plan. What fraction of her wages are taken out for these purposes?

Fractions Skills Practice

Circle the letter for the correct answer to each problem. Reduce all fractions to simplest terms.

1 $\dfrac{5}{6}$
$-\dfrac{1}{2}$

  A $\dfrac{4}{6}$ **C** $\dfrac{1}{4}$

  B $\dfrac{1}{3}$ **D** $\dfrac{1}{12}$

  E None of these

2 $1\dfrac{1}{3} \times \dfrac{2}{5} = $ _____

  F $1\dfrac{2}{15}$ **H** $\dfrac{8}{15}$

  G $1\dfrac{3}{8}$ **J** $\dfrac{4}{15}$

  K None of these

3 $2\dfrac{5}{12}$
$+1\dfrac{2}{3}$

  A $3\dfrac{3}{4}$ **C** $3\dfrac{7}{15}$

  B $3\dfrac{7}{12}$ **D** $4\dfrac{1}{12}$

  E None of these

4 $2\dfrac{1}{4} \div \dfrac{3}{8} = $ _____

  F 6 **H** $4\dfrac{2}{3}$

  G $2\dfrac{2}{3}$ **J** $\dfrac{27}{32}$

  K None of these

5 $\dfrac{3}{5} + \dfrac{3}{20} + \dfrac{1}{2} = $ _____

  A $1\dfrac{1}{5}$ **C** $\dfrac{7}{20}$

  B $\dfrac{7}{27}$ **D** $1\dfrac{1}{3}$

  E None of these

6 Which of these changes would increase $\dfrac{1}{3}$ to a number greater than 1?

  F Add $\dfrac{1}{2}$.

  G Multiply by 2.

  H Divide by 2.

  J Add $\dfrac{4}{5}$.

7 Which of these fractions is in lowest terms?

  A $\dfrac{3}{15}$

  B $\dfrac{7}{21}$

  C $\dfrac{9}{11}$

  D $\dfrac{4}{30}$

8 Kara spends about 45 dollars a week on groceries. Which choice below shows how much she spends on groceries per day?

  F $\dfrac{7}{45}$

  G $\dfrac{45}{7}$

  H $\dfrac{1}{7} \div 45$

  J $7 \times \dfrac{1}{45}$

9 One-eighth of the population of Riverview is unemployed. The population is 52,000. How many residents of Riverview are unemployed?

  A 650

  B 1,423

  C 6,500

  D 4,160

10 $1\frac{3}{4}$ **F** $1\frac{1}{2}$ **H** $\frac{1}{4}$
   $-\frac{1}{2}$ **G** $1\frac{1}{4}$ **J** $1\frac{1}{8}$
   K None of these

11 $5\frac{1}{3} \times 3\frac{3}{4} = $ _____
   A $15\frac{1}{4}$ **C** $18\frac{3}{4}$
   B $15\frac{3}{7}$ **D** 20
   E None of these

12 $6\frac{1}{4} - 2\frac{1}{3} = $ _____
   F $4\frac{1}{4}$ **H** $4\frac{11}{12}$
   G $3\frac{11}{12}$ **J** $4\frac{3}{4}$
   K None of these

13 $9 \div \frac{3}{4} = $ _____
   A 12   **C** $\frac{1}{12}$
   B $\frac{3}{4}$ **D** $6\frac{3}{4}$
   E None of these

14 $\frac{1}{3} \times \frac{3}{5} \times \frac{5}{8} = $ _____
   F $\frac{1}{9}$ **H** $\frac{17}{120}$
   G $\frac{1}{5}$ **J** $1\frac{2}{3}$
   K None of these

15 $6\frac{1}{5} \div 2\frac{2}{5} = $ _____
   A $\frac{12}{31}$ **C** $14\frac{22}{24}$
   B $2\frac{7}{12}$ **D** $3\frac{1}{2}$
   E None of these

16 Which group of fractions is in order from least to greatest?

   F $\frac{1}{3}, \frac{3}{4}, \frac{1}{9}, \frac{5}{8}$

   G $\frac{1}{3}, \frac{1}{9}, \frac{5}{8}, \frac{3}{4}$

   H $\frac{1}{9}, \frac{1}{3}, \frac{5}{8}, \frac{3}{4}$

   J $\frac{1}{3}, \frac{3}{4}, \frac{5}{8}, \frac{1}{9}$

17 One-half of a carrot cake must be divided evenly among 6 children. What fraction of the cake will each child get?

   A $\frac{1}{8}$

   B $\frac{1}{6}$

   C $\frac{1}{3}$

   D $\frac{1}{12}$

18 In Springfield, $\frac{3}{5}$ of the adults have children and $\frac{1}{4}$ of those adults have children under the age of 5. Which of these number sentences could you use to find the fraction of adults in Springfield that have children under 5?

   F $\frac{3}{5} \times \frac{1}{4} = $ _____

   G $\frac{3}{5} \div \frac{1}{4} = $ _____

   H $\frac{3}{5} - \frac{1}{4} = $ _____

   J $\frac{3}{5} + \frac{1}{4} = $ _____

19 Which of these fractions is equal to $\frac{3}{4}$?

   A $\frac{9}{8}$ **C** $\frac{12}{15}$

   B $\frac{5}{6}$ **D** $\frac{15}{20}$

Fractions Skills Practice

Signed Numbers

Positive and Negative Numbers

Look at the number line below:

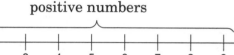

The numbers to the right of zero, such as 1, 2, $3\frac{1}{2}$, 5.7, and so on, are called the **positive numbers.** You can write the number positive five as "5" or "+5." The numbers to the left of zero, such as $-\frac{1}{2}$, –2, –8.5, and so on, are called the **negative numbers.** You always use the negative sign "–" to indicate a negative number.

You may already use negative numbers to refer to temperatures below zero. For example, two degrees below zero is written as –2°. You may also see negative numbers on bills, bank statements, and other balance sheets. The negative numbers represent money spent, while positive numbers stand for money earned.

As you move to the left on a number line, the values get smaller. Since –7 is to the left of –5, that means –7 is *smaller* than –5. Think about it this way: –7°F is colder than –5°F, or someone who owes 7 dollars has less money than someone who owes 5 dollars.

On a number line, we often want to know how far a number is from zero. A number's distance from zero is called its **absolute value.** The symbol for absolute value is $|\ |$.

The absolute value of –2 is 2 and of +5 is 5. In symbols: $|-2| = 2$ and $|5| = 5$.

PRACTICE

For problems 1–14, write a ">" symbol, a "<" symbol, or an "=" sign in each box to show whether the first number is greater than, less than, or equal to the second number.

1 –1 ☐ 0

2 0 ☐ 1

3 1 ☐ –1

4 –1 ☐ –2

5 –5 ☐ –3

6 $-\frac{1}{2}$ ☐ $-\frac{1}{4}$

7 35 ☐ –52

8 –1.3 ☐ –0.6

9 $|-4|$ ☐ 4

10 0 ☐ $|-5|$

11 –0.9 ☐ $|0.3|$

12 –0.03 ☐ –0.3

13 $-\frac{1}{5}$ ☐ $-\frac{1}{7}$

14 –0.2 ☐ –2

For problems 15–18, write the value of each expression.

15 $|-3| = $ ____

16 $|9| = $ ____

17 $|3| + |-3| = $ ____

18 $|25 - 11| = $ ____

19 Arrange these numbers from least to greatest: –2, 0, –5, 1

Adding Signed Numbers

You can use a number line to illustrate addition. Find the first number on the number line. Then think of addition as a matter of moving right or left on the number line. If you are adding a positive number, you move right. If you are adding a negative number, you move left.

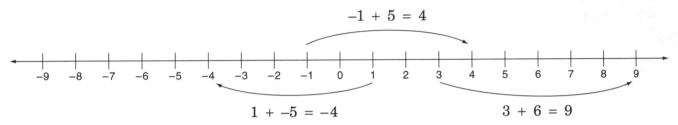

$$-1 + 5 = 4$$

$$1 + -5 = -4 \qquad 3 + 6 = 9$$

You can also use the absolute values of two signed numbers to find their sum. Here are the two rules.

If the numbers in an addition problem have the same sign (negative or positive), then: 1. Find the sum of their absolute values. 2. Give that sum their common sign. **Examples:** $\quad 2 + 2 = 4$ $\qquad\qquad -3 + -4 = -7$	If the numbers in an addition problem have different signs, then: 1. Find the difference of their absolute values. 2. Give that difference the sign of the number whose absolute value is larger. **Examples:** $\quad 2 + -2 = 0$ $\qquad\qquad -3 + 4 = 1$ $\qquad\qquad 4 + -5 = -1$

PRACTICE

Use a number line or the rules above to fill in each blank. *Hint:* **If you must add three or more numbers, add them two at a time, in any order.**

1 What number is 9 to the right of zero? _____

2 What number is 9 to the left of zero? _____

3 What is 3 to the right of −1 (add 3)? _____

4 What is 4 to the left of 2 (add −4)? _____

5 What is 5 to the left of −3 (add −5)? _____

6 To start at −3 and end at an integer greater than or equal to zero, you must move at least __?__ units to the right. _____

7 $0 + (-6) =$ _____

8 $1 + (-7) =$ _____

9 $-3 + 3 =$ _____

10 $-2 + (-2) =$ _____

11 $-12 + 8 =$ _____

12 $-10 + (-14) =$ _____

13 If you start at zero, what change will bring you to a number that is six less than zero?

A Add 6.
B Add −6.
C Add a number greater than −6.
D Add a number less than −6.

14 The temperature was −8°C. Then it rose 10 degrees. What was the temperature then?

Subtracting Signed Numbers

When you subtract one number from another, you are finding the difference between them. Therefore, you can solve a subtraction problem by finding the distance between two numbers on a number line. If you have moved left to solve the problem, give your answer a positive sign. If you have moved right, the difference is negative.

The distance from −4 to −9 is 5, so −4 − (−9) = 5.

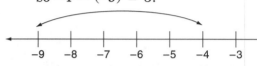

The distance from 4 to −1 is 5, so 4 − (−1) = 5.

Subtraction is the opposite, or inverse, of addition. You can think of subtraction as *adding the opposite of a number.*

Examples:

$$5 - 9 = 5 + (-9) = -4$$
$$-2 - (-4) = -2 + 4 = 2$$
$$-8 - (-5) = -8 + 5 = -3$$

Rules for Subtracting Signed Numbers

1. Change the sign of the number being subtracted.
2. Change the subtraction sign in the problem to an addition sign.
3. Follow the rules (on page 48) to add positive and negative numbers.

PRACTICE

Solve each problem below.

1 $5 - (-4) =$ _____

2 $-6 - 2 =$ _____

3 $-10 - (-5) =$ _____

4 $-1 - (-1) =$ _____

5 $12 - (-5) =$ _____

6 $20 - (-10) =$ _____

7 $-14 - (-12) =$ _____

8 $-5 - 5 =$ _____

9 $0 - (-12) =$ _____

10 Which of these changes will take you from −3 to 0?

   A Add −3.
   B Subtract 3.
   C Subtract −3.

11 If you start at −2, which of these changes would *decrease* the value?

   F Add −1.
   G Add 1.
   H Subtract −1.

12 Which of these operations has the same affect as the operation "subtract 5"?

   A Add −5.
   B Add 5.
   C Subtract −5.

Use this table for 13 and 14.

City	1990 Pop.	Change
Centralia	52,000	+12,400
Oseola	12,500	+2,400
Springfield	31,900	−1,400
Ryan	18,700	−200

13 What is the current population of Springfield?

14 Which is correct?

   A There are 200 more people living in Ryan now than there were in 1990.
   B There are 200 fewer people living in Ryan now than there were in 1990.

Multiplying and Dividing Signed Numbers

If you multiply a negative number by a positive number, the product is negative. Here is an example: If you owe a credit card company one hundred dollars, your balance is –$100.00. If your brother owes three times as much (3 × –$100.00), his balance is –$300. So:

$$3 \times -100 = -300$$

In a division problem, if one number is negative and the other is positive, then the quotient is negative. For example, if you owe 100 and your sister owes one-fourth as much, she owes –$25. So:

$$-100 \div 4 = \frac{-100}{4} = -25$$

Here are rules for multiplying and dividing signed numbers:

> To multiply signed numbers, multiply their absolute values.
> To, divide signed numbers, divide their absolute values.
>
> Then:
> If the two numbers have the same sign, write the answer as a positive number.
> If the two numbers have different signs, write the answer as a negative number.

Here are rules for simplifying fractions with signed numbers.

> If the numerator and the denominator have the *same sign,* the fraction is positive.
> $$\frac{-10}{-20} = \frac{1}{2} \qquad \frac{5}{25} = \frac{1}{5}$$
> If the numerator and the denominator have *different signs,* the fraction is negative.
> $$\frac{3}{-9} = -\frac{1}{3} \qquad \frac{-5}{20} = -\frac{1}{4}$$

PRACTICE

Solve each problem below.

1 $5 \times (-5) =$ _____

2 $-3 \times (-2) =$ _____

3 $-4 \times 4 =$ _____

4 $10 \times (-3) =$ _____

5 $16 \times (-2) =$ _____

6 $-10 \times (-10) =$ _____

7 $-8 \times (-5) =$ _____

8 $-2 \times 6 =$ _____

9 $10 \times (-12) =$ _____

10 $18 \div (-9) =$ _____

11 $40 \div (-10) =$ _____

12 $-60 \div (-3) =$ _____

13 $-75 \div 25 =$ _____

14 $-100 \div (-10) =$ _____

15 $240 \div (-2) =$ _____

16 $36 \div (-3) =$ _____

Simplify each fraction.

17 $\frac{-18}{-90} =$ _____

18 $\frac{40}{-88} =$ _____

19 $\frac{-15}{60} =$ _____

20 $\frac{-12}{-42} =$ _____

21 $\frac{-12}{48} =$ _____

Multiplying and Dividing Signed Numbers

Solving Mixed Word Problems

The problems below involve addition, subtraction, multiplication, or division of signed numbers. Set up and solve each problem. Use a negative sign ("–") for negative values. *Hint:* Some of these are two-step problems.

This Dominic's credit card statement for July. Use it to answer questions 1 through 4.

Previous balance	–$62.91
Transactions:	
July 5	–19.50
July 12	–30.10
July 14	+15.00
July 22	–52.87
Total transactions:	**–87.47**
Fees and/or credits:	**–6.75**

1   At the beginning of the month, did Dominic owe money or did he have a credit?   _____

2   On what date, if any, did the credit card company receive a payment from Dominic?   _____

3   The last entry in the statement is a total of all the fees and bonuses the credit card company gave Dominic. Overall, did the company give money to Dominic or charge him money?   _____

4   What was Dominic's balance (or total) at the end of July?   _____

5   The temperature is at –2°F, and it is dropping 1 degree every 10 minutes. If this keeps up, what will the temperature be in an hour (60 minutes)?

6   The balance in Jordan's bank account is $200.00. Every month, the bank adds –$15.00 to the account. If Jordan doesn't put any money in the account or take any out, what will his balance be at the end of a year (12 months)?

7   A scientist drops an instrument into the ocean. It drops 2 feet every second. How long will it take the instrument to reach a depth of –100 feet?

Signed Numbers Skills Practice

Circle the letter for the correct answer to each problem.

1 Reduce $\frac{-6}{-6}$.

A 1
B –1
C 0
D $-\frac{1}{6}$
E None of these

2 $60 + (-15) = $ _____

F 75
G –75
H 45
J –45
K None of these

3 $-15 \div (-5) = $ _____

A 3
B –3
C 75
D –75
E None of these

4 $-12 \times 3 = $ _____

F 36
G 4
H –36
J –4
K None of these

5 $10 + (-2) + (-2) = $ _____

A 10
B 8
C 6
D –6
E None of these

6 $\mid 2 - (-1) \mid = $ _____

F 3
G –3
H 1
J –1
K None of these

7 Which group of numbers is in order from least to greatest?

A $0, -1, -2, 5, 9$
B $-1, -2, 0, 5, 9$
C $-2, -1, 0, 5, 9$
D $0, -2, -1, 5, 9$

8 Which of these number sentences is true?

F $2 < -3$
G $-3 < -2$
H $0 < -3$
J $-3 < -4$

9 Which of the following choices, if it is added to 5, creates a sum that is less than 0?

A any number less than 5
B any number less than 0
C any number less than –5
D None of these

10 The town of Sedalia is 800 feet below sea level. That is an altitude of –800 feet. There is a quarry just outside of town that is 450 feet deep. What is the altitude at the bottom of the quarry?

- **F** 350 feet
- **G** –350 feet
- **H** 1,250 feet
- **J** –1,250 feet

11 Which of these expressions has the same value as –2 – (–4)?

- **A** 2 + 4
- **B** –2 + 4
- **C** –2 + (–4)
- **D** 2 + (–4)

12 What numbers can be added to –2 to get a number greater than 0?

- **F** any number less than 2
- **G** any number less than –2
- **H** any number greater than 2
- **J** any number greater than –2

13 Which of these expressions has the same value as –3 × (–5)?

- **A** 3 ÷ 5
- **B** 3 × (–5)
- **C** $\frac{1}{3} \times \frac{1}{5}$
- **D** 3 × 5

14 Which of these numbers is between –2 and –8?

- **F** 0
- **G** –1
- **H** 3
- **J** –5

Use the following information to do Numbers 15 through 17.

Carlos is a self-employed carpenter. This balance sheet shows his expenses and income for two recent projects.

Hiller House	
Lumber	–$1,800.00
Hardware	–60.00
Fee	+2,700.00
Smith House	
Lumber	–$2,200.00
Hardware	–260.00
Crane Rental	–360.00
Fee	+4,200.00

15 What were Carlos' total expenses on the Smith House?

- **A** $2,820.00
- **B** $2,460.00
- **C** $7,020.00
- **D** $1,380.00

16 What is the profit (income – expenses) for the Hiller House?

- **F** $840.00
- **G** $900.00
- **H** $960.00
- **J** $4,560.00

17 Carlos spends $800.00 a month on insurance. Also, last month he paid an engineer $1,500.00 for help on a project. How should he enter these figures in his balance sheet?

- **A** as –$800.00 and +$1,500.00
- **B** as –$800.00 and –$1,500.00
- **C** as –$800.00 and +$1,500.00
- **D** as +$800.00 and –$1,500.00

Ratio and Percent

Writing Ratios

Each of the following five statements is a *ratio:*

There are 4 men *for every* woman in the group.

There is 1 murder committed *every* 5 minutes.

He scores one free throw *out of every* three he attempts.

The chocolates cost $8.00 *per* pound.

A **ratio** is a comparison of two numbers. Ratios can be written in three ways:

1 to 2   **1 : 2** $\frac{1}{2}$

The four ratios from the top of this page are shown below. Each is written in fraction form and with a colon. Notice that the word *per* refers to "1 unit."

$\frac{4 \text{ men}}{1 \text{ woman}}$	$\frac{1 \text{ murder}}{5 \text{ minutes}}$	$\frac{1 \text{ score}}{3 \text{ attempts}}$	$\frac{\$8.00}{1 \text{ pound}}$
4 men : 1 woman	1 murder : 5 minutes	1 score : 3 attempts	$8.00 : 1 pound

All ratios should be written as fractions reduced to lowest terms. If a ratio is an improper fraction, *do not* write it as a mixed number.

PRACTICE

Write a ratio for each relationship below. Be careful to set each ratio up in the order indicated. (*Hint:* You will have to add or subtract to find some of the numbers you need.)

Write each ratio with a colon.

1   Zack spent $150.00 for 3 tickets. What was the ratio of money to tickets? _____

2   In a local election, 1,290 people voted for Beeler and 2,580 people voted for Price. What was the ratio of votes for Beeler to votes for Price? _____

Write each ratio as a fraction. Simplify whenever possible.

3   There are 24 cans in every carton of soup. Write the ratio of one can to the total number of cans in the carton. _____

4   The team won 18 games out of the 27 they played. Write the ratio of wins to losses. _____

5   You get 25 pesos for every 5 dollars. Write the ratio of pesos to dollars. _____

Writing Proportions

If you write an equation with one ratio equal to another ratio, such as $\frac{3}{6} = \frac{6}{12}$, you have a **proportion.** When you think of ratios as fractions, then the two ratios in a proportion are **equivalent fractions.**

Suppose that 4 gallons of rain come into a basement during a 10-minute storm. At the right are some ratios of amounts of rain to length of the storm.

$$\frac{\text{amount of rain}}{\text{length of storm}} =$$

$$\frac{6 \text{ gallons}}{15 \text{ minutes}} = \frac{8 \text{ gal}}{20 \text{ min}} = \frac{12 \text{ gal}}{30 \text{ min}} = \frac{24 \text{ gal}}{1 \text{ hr}} = \frac{48 \text{ gal}}{2 \text{ hr}}$$

In this problem you know three of the numbers in a proportion. You need to find the fourth number.

Problem:

If 4 gallons of water come into a basement during a 10-minute storm, how much water will come into the basement during a 45-minute storm?

Solution:

Use ratios. Write the amounts of water in both numerators and the lengths of time in both denominators.

$$\frac{4 \text{ gal}}{10 \text{ min}} = \frac{? \text{ gal}}{45 \text{ min}}$$

Find a ratio equivalent to $\frac{4}{10}$ that has 45 in the denominator:

The amount of water that will come into the basement in 45 minutes is 18 gallons.

$$\frac{4}{10} = \frac{4 \times 4.5}{10 \times 4.5} = \frac{18}{45}$$

PRACTICE

For problems 1–3, write each proportion in fraction form. *Remember:* **Use like labels in the numerators and denominators.**

1   You need 2 eggs to make 12 muffins. How many eggs do you need to make 30 muffins?

2   The plumber charges $60.00 for 20 minutes of work. At that rate, how much would he charge for a 90-minute job?

3   The cost of 14 ounces of coconut milk is $2.50. How much would a 21-ounce can of coconut milk cost?

For problems 4–7, tell how to solve each problem.

4   $\dfrac{112 \text{ miles}}{3 \text{ hours}} = \dfrac{? \text{ miles}}{9 \text{ hours}}$

To solve, multiply 112 miles by _____.

5   $\dfrac{2 \text{ inches}}{5 \text{ feet}} = \dfrac{18 \text{ inches}}{? \text{ feet}}$

To solve, multiply 5 feet by _____.

6   $\dfrac{130 \text{ baskets}}{30 \text{ days}} = \dfrac{? \text{ baskets}}{10 \text{ days}}$

To solve, divide 130 baskets by _____.

7   $\dfrac{6 \text{ pounds}}{\$15.00} = \dfrac{18 \text{ pounds}}{?}$

To solve, multiply $15.00 by _____.

When two fractions or ratios are equal, you can write a proportion. An example is $\frac{2}{3} = \frac{8}{12}$.

Below, notice that if you **cross multiply** by finding the products 3×8 and 2×12, the two products are equal.

Proportion	**Cross Multiply**	**Cross Multiply**
$\frac{2}{3} = \frac{8}{12}$	$\frac{2}{3} \diagdown\diagup \frac{8}{12}$	$3 \times 8 = 2 \times 12$
		$24 = 24$

The result is true for any proportion: When you cross multiply in a proportion, the products are equal. Therefore, you can solve a proportion by following these steps:

1. Cross multiply the numbers in the proportion.

    $$\frac{5 \text{ fruit bars}}{\$2} = \frac{? \text{ fruit bars}}{\$10} \quad \text{Here is a proportion.}$$

    $$\$2 \times ? = \$50 \quad \text{Cross multiply.}$$

2. Divide to find the value of the "?" symbol.

    $$? = \frac{50}{2} = \$25 \quad \text{Divide to get "?" alone.}$$

PRACTICE

Cross multiply to solve each problem below. Round all answers to the nearest hundredth.

8   $\dfrac{1 \text{ mile}}{5280 \text{ feet}} = \dfrac{1.5 \text{ miles}}{? \text{ feet}}$

9   $\dfrac{3 \text{ flowers}}{\$4.00} = \dfrac{? \text{ flowers}}{\$28.00}$

10   $\dfrac{1 \text{ inch}}{2.54 \text{ centimeters}} = \dfrac{24 \text{ inches}}{? \text{ centimeters}}$

11   $\dfrac{3 \text{ feet}}{2.5 \text{ hours}} = \dfrac{18 \text{ feet}}{? \text{ hours}}$

12   It takes you 10 minutes to walk 0.5 miles. At that rate, how long would it take to walk five miles?

13   It takes Jenna three months to save $275. At that rate, how much will she save in a year?

14   A basketball player makes 3 free throws out of every 5 he attempts. At that rate, how many free throws will he make in 42 attempts?

Writing Proportions

Percent

The word **percent** means "per hundred." A percent is a fraction with a written or unwritten denominator of 100.

Examples: $25\% = \frac{25}{100}$ or $\frac{1}{4}$ $50\% = \frac{50}{100}$ or $\frac{1}{2}$

To write a percent as a fraction, write the percent over 100. Then simplify the fraction.

$$30\% = \frac{30}{100} = \frac{3}{10} \times \frac{10}{10} = \frac{3}{10}$$

To write a fraction as a percent, rewrite the fraction so it has a denominator of 100. Then write the numerator with a "%" symbol.

$$\frac{4}{5} = \frac{4}{5} \times \frac{20}{20} = \frac{80}{100} = 80\%$$

To write a percent as a decimal, move the decimal point two places to the left.

$25\% = 0.25$
$50\% = 0.50$

To write a decimal as a percent, move the decimal point two places to the right.

$0.47 = 47\%$
$0.015 = 1.5\%$

PRACTICE

Fill in each blank below.

1 One whole is _____ %.

2 Circle each percent below that is more than one whole.

   30%   130% $1\frac{1}{2}\%$

   56%   650%   79%

   211%   40.5%   65%

3 What percent is $\frac{1}{2}$? _____

4 What percent is $\frac{1}{4}$? _____

5 Which is greater, $\frac{1}{2}\%$ or 0.5? _____

6 Rewrite these percents in order from least to greatest:
   $\frac{1}{4}\%$, $50\frac{1}{2}\%$, 50%, 150%

7 If there are 100 candies in a box, and 40% are chocolates, how many chocolate candies are there?

Write each percent below as a decimal number.

8 13%   11 $4\frac{1}{2}\%$

9 5%   12 75%

10 115%   13 0.6%

Write each decimal number below as a percent.

14 0.3   17 1.9

15 0.02   18 40.2

16 0.15   19 0.12

If a percent is written as a mixed number, the method below shows how to write it as a fraction.

Problem: Write $33\frac{1}{3}\%$ as a fraction.

Solution:

Start by writing the mixed number as an improper fraction.

$$33\frac{1}{3}\% = \frac{33}{1} + \frac{1}{3}$$
$$= \frac{99}{3} + \frac{1}{3} = \frac{100}{3}\%$$

Then write the percent as a fraction by dividing by 100 and dropping the "%" sign.

$$\frac{100}{3}\% = \frac{100}{3} \div 100$$
$$= \frac{100}{3} \times \frac{1}{100} = \frac{1}{3}$$

Complete the table below. Reduce all fractions to simplest terms. *Hint:* **You will use the values in these tables over and over again, so it would be worthwhile to memorize them.**

20

Percent	Fraction	Decimal
5%		
	$\frac{1}{10}$	
		0.15
20%		
	$\frac{1}{4}$	
		0.3
$33\frac{1}{2}\%$		
		0.4
$44\frac{3}{4}\%$		
50%		
$57\frac{1}{4}\%$	$\frac{229}{400}$	
	$\frac{3}{5}$	
		0.72
		0.75
80%		
	$\frac{9}{10}$	

Problem:
Write $\frac{1}{22}$ as a percent.

Solution:
Divide:

```
        0.04545
  22)1.00000
       88
      ----
      120
      110
      ----
       100
        88
       ----
       120
       110
       ----
        10
```

Round the quotient to the thousandths place to get 0.045. Then move the decimal point two places to the right, and write the "%" sign, to get 4.5%.

Answer: $\frac{1}{22} = 4.5\%$

Use division to complete the table below.

21

Fraction	Percent	Fraction	Percent
$\frac{1}{3}$		$\frac{2}{3}$	
$\frac{1}{12}$		$\frac{8}{9}$	
$\frac{5}{7}$		$\frac{1}{8}$	

Finding a Percent of a Number

To find a percent of a given number, follow these steps:

1. Write the percent as a decimal or as a fraction.
2. Multiply that value by the given number.

Problem: A $75.00 blanket is marked 20% off. What is 20% of $75.00?

Solution 1:
Write the percent as a decimal.

Multiply that decimal by the given number.

$$20\% = 0.20$$

$$
\begin{array}{r}
\$75.00 \\
\times\ 0.20 \\
\hline
15.0000 \ or \ \$15.00
\end{array}
$$

Solution 2:
Write the percent as a fraction.

Multiply that fraction by the given number.

$$20\% = \frac{20}{100} = \frac{1}{5}$$

$$\frac{1}{5} \times 75 = \frac{1}{5} \times \frac{75}{1} = 15 \text{ or } \$15$$

Remember: When your answer is in dollars, make sure it has the correct number of decimal places. When you answer is *not* in dollars, you can write or erase zeros at the right of a decimal number without changing its value.

PRACTICE

Find the indicated percent of each number. Round all answers to the nearest hundredth.

Use the information below to do numbers 8 and 9.

1. What is 40% of 520?

2. What is 21% of 17?

3. What is 15.5% of $604.00?

4. What is $5\frac{1}{2}\%$ of 400?

5. What is $33\frac{1}{3}\%$ of $9.00?

6. The sales tax in town is $9\frac{1}{2}\%$. How much will you pay for a lamp marked $65.00? (Hint: This is a 2-step problem.)

7. A $40.00 shirt is marked down 28%. How much does it cost? (*Hint:* This is a two-step problem.)

There are 51,700 registered voters living in Broken Bow. The city council expects 40% of voters to participate in the next election. To win that election, a candidate must get 51% of the votes cast.

8. If 34,000 people vote, how many votes must the winner get?

9. If the city council's prediction is correct, how many votes must the winner get?

Finding What Percent One Number Is of Another

To find what percent one number is of another, put the numbers in fraction form. In the fraction, write the "part" in the numerator and the "whole" in the denominator. Then reduce the fraction and write it as a percent.

Problem:

Quinn spends 6 hours a week of his free time at a gym. He has only 40 hours of free time each week. What percent of his free time does Quinn spend at the gym?

Solution:

First, write 6 and 40 as a fraction. $\dfrac{6}{40}$

Simplify the fraction. $\dfrac{6}{40} = \dfrac{2}{2} \times \dfrac{3}{20} = \dfrac{3}{20}$

Write the fraction as a percent by rewriting the fraction with a denominator of 100. $\dfrac{3}{20} = \dfrac{3}{20} \times \dfrac{5}{5} = \dfrac{15}{100} = 15\%$

Answer: Quinn spends 15% of his free time at the gym.

PRACTICE

Solve each problem below. Round all answers to the nearest percent.

1 What percent of 50 is 12?

2 What percent of $5.50 is $0.22?

3 What percent of 60 is 84? (Notice that 84 is greater than 60.)

4 What percent of 130 is 72.8?

5 A $120.00 humidifier has been marked down to $90.00. By what percent has it been marked down? (This problem has two steps.)

6 The sticker price on the new car Quang wants is $24,500. The car dealer offers it to him for $22,050. What percent of the sticker price is he willing to take off? (*Hint:* This problem has two steps.)

7 You have driven 75 miles of a 375-mile trip. What percent of the trip is left? (This problem has two steps.)

8 Kyla makes 150 sales calls. Of the people she calls, 25 people place orders. What percent of her calls were successful?

Finding the Total When a Percent Is Given

In the problem below, the part and the percent are given. To find the total number of people, follow these steps:

1.   Write the percent as a fraction.
2.   Divide the given number by that fraction.

Problem: At Unified Systems, 265 workers are on strike. That is 35% of the total number of workers at the factory. How many workers are there at Unified Systems?

Solution:

First, write 35% as a fraction.

$$\frac{35}{100}$$

Simplify the fraction.

$$\frac{35}{100} = \frac{5}{5} \times \frac{7}{20} = \frac{7}{20}$$

Divide 265 by the fraction.

$$265 \div \frac{7}{20} = \frac{265}{1} \times \frac{20}{7} = \frac{5300}{7} = 757.1428\ldots$$

The answer refers to a number of people, so round the answer to the nearest whole number. The total number of workers at Unified Systems is 757.

PRACTICE

Solve each problem below. Round each answer to the nearest whole number.

1   Twenty-one is 60% of what number?

2   Four and eight tenths is 32 % of what number?

3   Three hundred is 120% of what number?

4   Fifty-two is 25% of what number?

5   The accountant for Lilly's Gifts reported that the store sold $69,000 worth of merchandise this December. The accountant said that amount was 43% of the total sales for the year. How much merchandise did the store sell this year? (Round your answer to the nearest thousand dollars.)

6   Rod got a statement from the bank saying he had paid off 90% of his car loan. It said that so far he had paid $19,260.00. How much was the original loan?

7   In 1997, the number of American schools that had modems was 41 million. That was 47.7% of all American schools. About how many schools were there in America that year? (Round your answer to the nearest million.)

Mixed Practice with Percent

Solve each problem below.

1 In a poll, 86 percent of the 2,000 parents questioned said that they think their children are "good kids." How many parents gave this answer?

2 A couch originally marked $750 is on sale for 20% off. How much does it cost now?

3 Celine bought a coat for $120.00 that was originally marked $160.00. What percent of the original price did she pay?

4 Ten years ago, an antique Jenny Lynn plate sold for $12.00. Today it sells for $66.00. Today's price is what percent of the old price?

This list shows the interest rates a bank is offering to people who invest in certificates of deposit (CDS). Use the list to do problems 5 through 7.

Certificates of Deposit

1-year CD: 4% yearly interest
2-year CD: 6% yearly interest
3-year CD: 8% yearly interest

$50.00 penalty for early withdrawal

---

5 If you put $2,400 in a 1-year CD, how much money will you have at the end of one year?

6 If you put that $2,400 in a 2-year CD, how much interest will you make by the time it matures?

7 Roger puts $3,000 in a 2-year CD, but he withdraws it after just one year. How much did he make on his investment?

Ratio and Percent Skills Practice

Circle the letter for the correct answer to each problem. Reduce all fractions to simplest terms.

1   10.5% of $300.00 = _____

    A $31.50
    B $315.00
    C $3.90
    D $3.50
    E   None of these

2

135% of ☐ = 270

    F   135
    G   500
    H   364.5
    J   200
    K   None of these

3

What percent of 120 is 20?

    A   67%
    B   60%
    C   40%
    D   21%
    E   None of these

4

$4\frac{1}{2}$% of $5.30 = __

    F $0.24
    G $0.45
    H $0.32
    J $0.12
    K   None of these

5

What percent of 440 is 264?

    A   50%
    B   60%
    C   65%
    D   70%
    E   None of these

6   Grant bought a computer game for $32.00 plus 8% tax. How much did he pay?

    F $37.60
    G $34.56
    H $34.50
    J $35.50

7   Candice just got a 6% salary increase. She was making $28,700 per year. How much will she now make each year?

    A $30,422
    B $17,220
    C $27,220
    D $19,978

8   Kelly buys a $50.00 shirt on sale for $45.00. How much was the shirt marked down?

    F   5%
    G   10%
    H   90%
    J   15%

9 Which of the following is equal to 60%?

A $\frac{2}{3}$ **C** $\frac{3}{5}$

B 0.06   **D** $\frac{6}{100}$

10 Ronnie buys a blouse for $65.00 plus 8% tax. She uses this number sentence to figure out how much she owes.

$65.00 × 0.08 = \rule{2cm}{0.4pt}$

What does the answer to the number sentence represent?

F the price of the blouse before tax
G the cost of the blouse, including tax
H what percentage $65.00 is of the final cost
J how much tax she will pay

11 Which of these number sentences could you use to solve the following proportion:

$\frac{35}{125} = \frac{90}{n}$

A $35 × 90 = 125 × n$
B $35 × n = 125 × 90$
C $125 × 35 = 90 × n$
D Any of these number sentences would work.

12 Akiko paid $325.00 for a chair that was originally marked $500.00. By what percent was the chair marked down?

F 35%
G 65%
H 0.35%
J 28%

13 If one dozen eggs cost 90 cents, how much will 20 eggs cost?

A $1.20
B $1.50
C $1.00
D $1.30

14 Mark earns $1,000 a month plus a 6% commission on all his sales. In August, his sales were a total of $12,000. How much money did he earn altogether that month?

F $720.00
G $780.00
H $1,720.00
J $8,200.00

15 Tuition at the Performers' Dance School is $150.00 per 20-week semester. When Rosa started class, there were only 8 weeks left in the winter semester. Which of these proportions could she use to figure out how much tuition she owes?

A $\frac{\$150.00}{20} = \frac{8}{x}$

B $\frac{20}{\$150.00} = \frac{x}{8}$

C $\frac{\$150.00}{20} = \frac{x}{8}$

D Any of these proportions would work.

Ratio and Percent Skills Practice

Data Interpretation

Reading a Table

Tables and graphs are useful ways to organize information and show many numbers. In order to understand a table or a graph, always begin by reading the title and the headings. They explain the relationships shown in a table or graph.

To find out who will be waiting tables on Tuesday, start by looking along the row for "Waiters."

Then look down the column for Tuesday. There is one box (or "cell") that appears in that column and that row. The people waiting tables on Tuesday are Adolfo, Lana, and Alicia.

	Sun.	Tues.	Wed.	Thurs.	Fri.	Sat.
Cook	Marcus Allen	Marcus	Stan	Marcus	Marcus Allen	Stan
Bus staff	Stan			Stan	Stan	Adolfo
Waiters	Alicia Adolfo Connie Greta	Adolfo Lana Alicia	Adolfo Allen Lana	Alicia Greta Lana	Alicia Adolfo Connie	Connie Allen
Host	Lana				Lana	Lana

PRACTICE

Use the table above to answer each question.

1 Who will be cooking Thursday morning? _____

2 Who will be busing tables Saturday morning? _____

3 What job will Allen be doing Friday morning? _____

4 How many people will be working at the restaurant Wednesday morning? _____

5 On which of the mornings listed will the most people be working at the restaurant? _____

6 How many days will Connie work this week? _____

7 What day will the restaurant be closed? _____

8 On what day will both Greta and Allen be working? _____

Reading a Bar Graph

A bar graph uses bars to represent values. A longer bar represents a larger number.

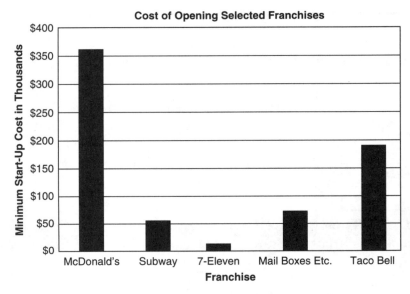

Cost of Opening Selected Franchises

Minimum Start-Up Cost in Thousands

Franchise

Question: How much does it cost to open a Subway franchise?

Solution: Use the labels along the bottom of the graph to find the bar for Subway. Using the length of the bar, find the number that corresponds to the top of the bar. It is a little more than $50.

Then find the units for the graph. The units are "thousands (of dollars)," so the answer is "A little over $50,000."

If the length of a bar falls between two numbers, estimate where it lies between the two numbers. For example, it might be halfway between the numbers, $\frac{1}{3}$ of the way, and so on. Then figure out the value for the end of the bar. For example, the top of the bar for Subway lies about one-tenth of the way from 50 to 100, which is a difference of 50. Since one-tenth of 50 is 5, the top of the bar is at 55, which represents $55,000. A good estimate for the cost to open a Subway franchise is $55,000.

PRACTICE

Estimate the minimum start-up cost for opening each of the franchises below.

1   McDonald's   _____

2   7-Eleven   _____

3   Mail Boxes Etc.   _____

4   Taco Bell   _____

Reading a Line Graph

A line graph uses points or dots to show values. The numbers along one side of the graph show the values of the points on the graph.

On the graph, lines between the points show whether the values are rising or falling. So line graphs show trends and changes in amounts.

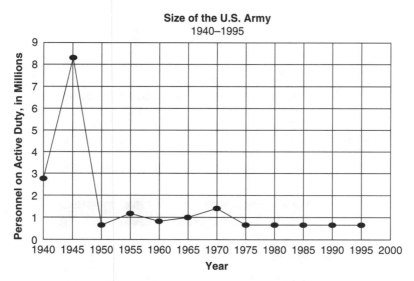

Question 1: What was the size of the U.S. Army in 1980?
Solution:
Find the label "1980" at the bottom of the graph.
Find the point on the graph that represents "1980."
Find the value for that point. It is about $\frac{4}{5}$ of the way from 0 to 1, and the side of the graph says that the numbers are "in millions."

Since $\frac{4}{5} \times 1,000,000 = 800,000$, the size of the U.S. Army in 1980 was about 800,000.

Question 2: Is the size of the U.S. Army rising or falling?

Solution:
The size has been pretty steady since 1950.

PRACTICE

Use the graph above to answer these questions.

1   About what was the size of the U.S. Army in 1940? _____

2   In which four of the years labeled did the army have over 1 million people? _____, _____

    _____, _____

3   What was the largest size of the army? _____

4   During which 20-year period did the size of the army stay most steady (unchanged)? _____

5   What year is represented by the last point on the graph? _____

6   Why was the army so much bigger during the 1940s than in any other period shown?

    _____

Using Numbers in a Graph

You can use the numbers in a graph to make comparisons. Start by finding each number. Then:
- ◆ You can find the **difference** between two numbers by subtracting.
- ◆ You can see how **many times larger** one number is than another by dividing.
- ◆ You can find **what fraction** one number is of another by forming a ratio.

PRACTICE

Use the graph below to answer questions 1 through 6

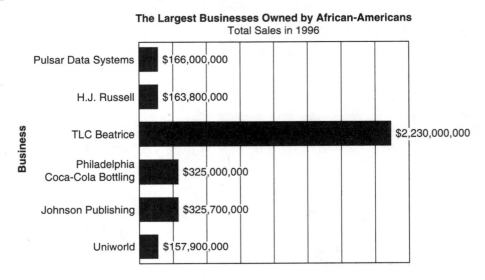

The Largest Businesses Owned by African-Americans
Total Sales in 1996

Pulsar Data Systems — $166,000,000
H.J. Russell — $163,800,000
TLC Beatrice — $2,230,000,000
Philadelphia Coca-Cola Bottling — $325,000,000
Johnson Publishing — $325,700,000
Uniworld — $157,900,000

1 In 1996, how much more money did H. J. Russell make than Uniworld? _____

2 TLC Beatrice is about _?_ times larger than the next largest African-American-owned business. (Round your answer to the nearest whole number.) _____

3 TLC Beatrice is about _?_ times larger than Uniworld. _____

4 H. J. Russell is about what fraction of the size of Johnson Publishing?

A $\frac{1}{2}$ C $\frac{1}{3}$

B $\frac{1}{4}$ D $\frac{1}{5}$

5 In 1996, how much more money did Philadelphia Coca-Cola Bottling make than Pulsar Data Systems?

_____

6 What is the second largest African-American-owned business in America?

_____

Using Numbers in a Graph

Finding the Percent of Change

To find the percent by which a value has increased or decreased, start by finding the amount of change.

Then divide that change by the original value, and express your answer as a percent.

Question: In the graph below, by what percent did the number of foreign-born U.S. residents increase between 1940 and 1990?

Solution:

Estimate values for 1940 and 1990, and subtract them.	20.0 million
	− 11.8 million
	8.2 million

The "original value" is the 1940-value, or 11.8 million, Divide 8.2 by 11.8, and write the result as a percent. There was a 69.4% increase.

$$8.2 \div 11.8 = 0.694$$

$$0.694 = 69.4\%$$

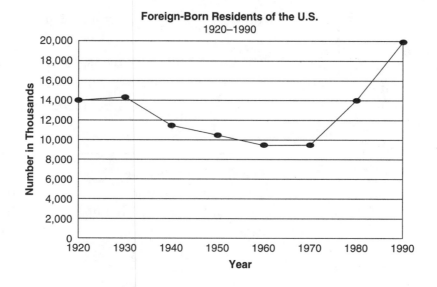

Foreign-Born Residents of the U.S.
1920–1990

PRACTICE

Use the graph above to answer these questions.

1 By what percent did the foreign-born population of the U.S. increase between 1980 and 1990?

_____

2 In what 10-year period did the foreign-born population *decrease* the most?

_____ to _____

3 Between 1930 and 1970, by what percent did the foreign-born population of the U.S. decrease?

_____

4 Between 1990 and 1996, the foreign-born population of the U.S. increased by 9.6 percent. What was the foreign-born population in 1996?

_____

5 There were about __?__ times as many foreign-born people in the U.S. in 1920 as there had been in 1970.

_____

Finding the Mean, Median, and Mode

There are several different meanings for "the most typical value" in a list of values.

◆ One meaning of "most typical value" is the **mean** or **average.** To find an average, add all the numbers in a set. Then divide that sum by the total number of values.

◆ The **median** in an ordered list of numbers is the middle value. If a list has an even number of values, then the median is the number *halfway between* the two middle numbers.

◆ The **mode** in a set of numbers is the value that appears most often. If no number appears more often than any other, then the set of values has "no mode."

Example: Danielle is taking a class for data-entry clerks. This list shows how long it took her to enter the data in several different documents.

	Time	Pages
Document 1	12 minutes	20
Document 2	23 minutes	55
Document 3	15 minutes	25
Document 4	8 minutes	20
Document 5	7 minutes	5
Totals:	**65 minutes**	**125**

The mean time *per document* is the total time divided by the number of documents:

$$\frac{65 \text{ minutes}}{5 \text{ documents}} = 13 \text{ minutes per document}$$

The mean time *per page* is the total time divided by the number of pages:

$$\frac{65 \text{ minutes}}{125 \text{ pages}} = 0.52 \text{ minutes per page}$$

The median time per document is the middle value in the ordered list of times:

7   8   12   15   23

↑
median time per document

No score appears more than once in the list, so there is no mode.

PRACTICE

Danielle's data-entry class took several timed tests using 12-page documents. This table shows the results of the tests. Fill in the missing numbers in the shaded section of the table, rounding all answers to the nearest tenth of a minute. Some students missed trials, so make sure you use the total number of documents and pages each student actually typed.

Student	Minutes Spent on Each 12-Page Document						Median Time Per Document	Mode	Mean Time Per Document	
	1	2	3	4	5	6			Per Doc.	Per Page
Lynn	5	7	8	9	10	12	8.5	None	8.5	0.7
Larry	10	12	14	10	X	X				
Roberto	8	6	7	10	12	X				
Tim	18	16	14	14	15	13				

Trends and Predictions

A pattern in a graph is called a **trend**.

Trends and Line Graphs

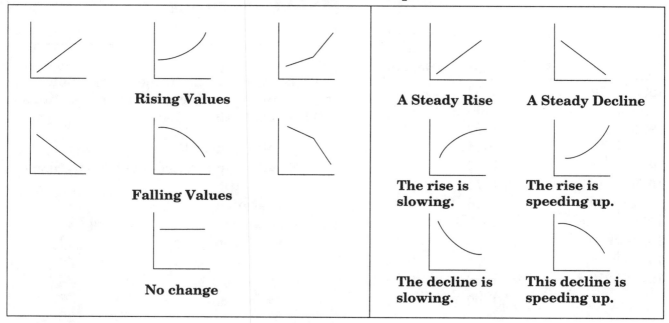

PRACTICE

Identify the trend, if any, that is shown in each graph. Then use the graphs to answer questions 1 through 4.

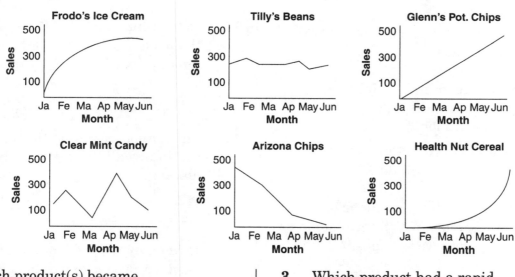

1 Which product(s) became more popular over the 6 months shown?

2 Which product(s) became much less popular?

3 Which product had a rapid rise in popularity followed by a leveling-off period?

4 Which product looks like it will be the biggest seller in July?

Look for patterns and trends in the table below. Then use it to answer questions 5 through 10.

Winning Times in the Summer Olympics 100-meter Race

Year	Winning Time (Seconds)	Year	Winning Time (Seconds)
1896	12	1952	10.4
1900	11	1956	10.5
1904	11	1960	10.2
1908	10.8	1964	10.0
1912	10.8	1968	9.95
1920	10.8	1972	10.14
1924	10.6	1976	10.06
1928	10.8	1980	10.25
1932	10.3	1984	9.99
1936	10.3	1988	9.92
1948	10.3	1992	9.96
		1996	9.84

5 Which of these trends appears in the winning times shown on the table?

   A Over time, the runners have become faster.

   B The early runners were faster than later runners.

   C Over the years, the runners have become older.

6 Given the trend in finishing times, which of these races must have been a big disappointment?

   F 1896   **H** 1900

   G 1968   **J** 1980

7 The Summer Olympics are held at regular intervals. Which three summer Olympic games were skipped?

8 What change is shown in the way this race was timed?

9 Which Olympic record in this race stood untied and unbeaten for the longest time?

10 Given the trends in this table, which of these is most likely to be the finishing time in this race in the year 2004?

   A 10.25 seconds
   B 9.04 seconds
   C 9.82 seconds

Look for patterns in the table below. Then use the table to answer questions 11 through 13.

Number of Rolls of Wallpaper Needed for Rooms of Different Sizes

Distance Around the Room	Wall Height		
	8 ft	10 ft	12 ft
36 ft	14	16	20
40 ft	14	18	22
44 ft	16	20	24
48 ft	18	22	26
52 ft	18	24	28
56 ft	20	26	30
60 ft	22	26	32
64 ft	22	28	34
68 ft	24	30	36

11 In the column labeled "Distance Around the Room," each value is the number directly above plus __?__.

12 Fill in another row on the table, continuing all the patterns that appear above it.

13 How many rolls of wallpaper would you need for a room 76 feet in perimeter with a wall height of 12 feet?

Reading a Circle Graph

A circle graph shows how a whole is divided into parts. It is divided into sections, like a pie, and each section stands for a fraction of the total. A larger section represents a larger fraction.

The labels in a circle graph are usually written as fractions or percents. To find the actual numbers for the sections of the graph, multiply each fraction or percent by the total number that is represented by the graph.

PRACTICE

Study each graph. Then use it to answer the questions below the graph.

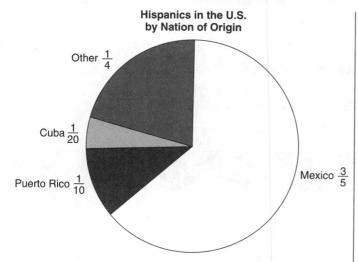

Hispanics in the U.S. by Nation of Origin

Other $\frac{1}{4}$

Cuba $\frac{1}{20}$

Puerto Rico $\frac{1}{10}$

Mexico $\frac{3}{5}$

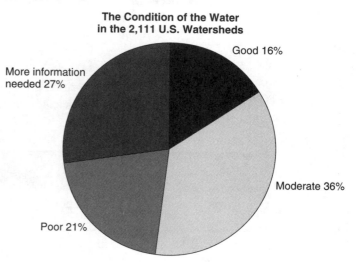

The Condition of the Water in the 2,111 U.S. Watersheds

Good 16%

More information needed 27%

Moderate 36%

Poor 21%

1 Twice as many Hispanic-Americans come from Puerto Rico as from __?__ . _____

2 What percent of Hispanic-Americans come from Mexico? _____

3 Together, people whose origins are Puerto Rico or Mexico are what fraction of the Hispanic-Americans living in the U.S.? _____

4 There are 22 million people of Hispanic origin living in the U.S. How many are people whose origins are Puerto Rico? _____

5 How many people of Mexican origin live in the U.S.? (*Hint:* Use the information in Question 4.) _____

6 Which fraction best describes how many U.S. watersheds have poor water?

A $\frac{1}{4}$ **C** $\frac{1}{20}$

B $\frac{1}{5}$ **D** $\frac{1}{8}$

7 In what fraction of U.S. watersheds has the water quality not been studied?

F almost $\frac{1}{2}$ **H** about $\frac{2}{3}$

G over $\frac{1}{4}$ **J** just $\frac{1}{10}$

8 To the nearest whole number, how many U.S. watersheds are known to have good water? _____

9 To the nearest whole number, how many U.S. watersheds are known to have poor water? _____

Reading a Complex Table or Graph

A bar graph may have several types of bars, a line graph may have several kinds of lines, or a table may have several sections. For these graphs and tables, carefully read all the labels. There may be a box, called a **key** or **legend,** explaining the labels.

PRACTICE

Use the table below to answer questions 1 through 5. Find the information you need in the table and check that it is correct *before* you do any figuring.

Schedule for the Week of May 6–12

MORNING SHIFT	Sun.	Tues.	Wed.	Thurs.	Fri.	Sat.
Cook	Marcus Allen	Marcus	Stan	Marcus	Marcus Allen	Stan
Bus Staff	Stan			Stan	Stan	Adolfo
Waiters	Alicia Adolfo Connie Greta	Adolfo Lana Alicia	Adolfo Allen Lana	Alicia Greta Lana	Alicia Adolfo Connie	Connie Allen
Host	Lana				Lana	Lana
DAY SHIFT	Sun.	Tues.	Wed.	Thurs.	Fri.	Sat.
Cook	Gail Tomas	Gail	Roger	Gail	Gail Tomas	Roger
Bus Staff	Roger			Roger	Roger	
Waiters	Wendy Mo	Sandy Wei	Sandy Wei	Wendy Mo Wei	Wendy Sandy Shawn	Shawn
Host	Wei				Wei	
EVENING SHIFT	Sun.	Tues.	Wed.	Thurs.	Fri.	Sat.
Cook	Dan S.	Dan S.	Bailey	Dan S.	Dan S.	Bailey
Bus Staff						
Waiters	Ali Juana	Juana Maria	Juana Maria	Ali Mary	Ali Juana	Dick Dan R.
Host	Maria				Maria	Maria

1 How many people will work the evening shift on Tuesday? _____

2 How many cooks does this restaurant employ? _____

3 What day and shift have the smallest staff? _____

4 Waiters make $3.50 an hour, and each shift is 8 hours long. How much will the day shift waiters be paid for their work on Thursday? _____

5 Cooks receive $72 per shift. How much will the restaurant pay its cooks for the work they do this week? _____

Use this graph to answer questions 6 through 15.

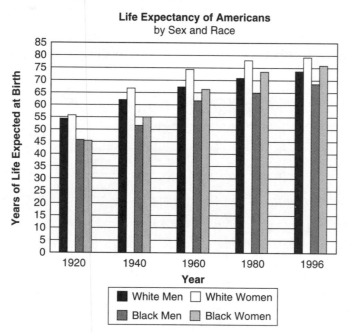

Life Expectancy of Americans
by Sex and Race

Years of Life Expected at Birth / Year

Legend: ■ White Men □ White Women ■ Black Men ■ Black Women

6 What is the life expectancy of Black Men born in 1940? _____

7 What is the life expectancy of White Women born in 1980? _____

8 Which group has the longest life expectancy shown on the table? _____

9 How much longer are White Women born in 1960 expected to live than White Men born that year? _____

10 How many years did the life expectancy of White Women rise between 1920 and 1996? _____

11 Overall, who lives longer, men or women? _____

12 What was the percent increase in the life expectancy of Black Women between 1920 and 1996? (Round your answer to the nearest whole percent.) _____

13 Has life expectancy risen more for White Men or Black Men as a group since 1940? _____

14 Which has *increased* the most, the difference between the life expectancies of Black Men and Black Women or the difference between the life expectancies of White Men and White Women? _____

15 Which of these is the best prediction of the life expectancy of White Men born in the year 2000?

   A 72 years **C** 76 years
   B 82 years **D** 89 years

Reading a Complex Table or Graph

Data Interpretation Skills Practice

This graph shows how many fatal airline crashes there were in each year from 1988 to 1996 and how many people died in them. Study the graph. Then do Numbers 1 through 5.

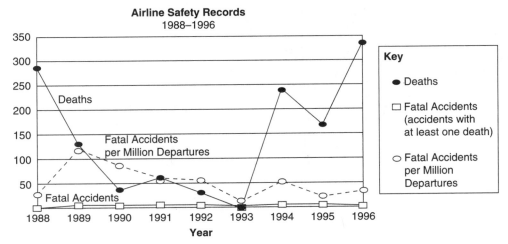

Airline Safety Records
1988–1996

1 In which of the following years were there the most deaths per accident?

A 1988
B 1990
C 1993
D 1994

2 By how much did the number of deaths drop between 1988 and 1990?

F about 250
G about 300
H about 350
J about 150

3 Suppose that the trend in fatal accidents per million departures continues from 1995 for the next five years. What is the best prediction for the number of fatal accidents per million departures in 1999?

A 50
B 60
C 30
D 2

4 In which of the following years was there the greatest increase from the previous year in the number of deaths?

F 1989
G 1992
H 1994
J 1996

5 Which of these statements best describes the trends shown in the graph?

A The number of fatal accidents and deaths tends to be rising while the number of fatal accidents per million is falling.

B While the number of deaths varies from year to year, the number of fatal accidents remains level, and the number of fatal accidents per million departures is generally falling.

C Airline fatalities will soon become a thing of the past.

D The number of deaths per accident is falling rapidly although the number of fatal accidents is rising.

Data Interpretation Skills Practice

These tables show a simplified version of income tax rates for 1997. (In the actual tables, each line after the first refers to the amount of money *more than* the previous category.) The figures are given as percents of the person's total taxable income. Study the tables. Then do Numbers 6 through 9.

Income Tax Rates

Single

Tax Rate	Taxable Income
15%	$0 to $24,650
28%	$24,561 to $59,750
31%	$59,751 to $124,650
36%	$124,651 to $271,050
39.6%	more than $271,050

Married Filing Jointly or Qualifying Widow(er)

Tax Rate	Taxable Income
15%	$0 to $41,200
28%	$41,201 to $99,600
31%	$99,601 to $151,750
36%	$151,751 to $271,050
39.6%	more than $271,050

Married Filing Separately

Tax Rate	Taxable Income
15%	$0 to $20,600
28%	$20,601 to $49,800
31%	$49,801 to $75,875
36%	$75,876 to $135,525
39.6%	more than $135,525

Head of Household

Tax Rate	Taxable Income
15%	$0 to $33,0500
28%	$33,051 to $85,350
31%	$85,351 to $138,200
36%	$138,201 to $271,050
39.6%	more than $271,050

6 The greatest increase in the tax rate occurs between which two income levels?

    F the first and second: 15% to 28%
    G the second and third: 28% and 31%
    H the third and fourth: 31% and 36%
    J the fourth and fifth: 36% and 39.6%

7 Rachelle is married, but she and her husband are filing separately. Rachelle earns $62,000 per year. According to these tables, how much must she pay in taxes?

    A $9,300
    B $19,220
    C $17,360
    D $22,320

8 A person who earns $140,000 a year will pay the most money in taxes if he or she is __?__ .

    F a single person
    G a married person filing separately
    H a married couple filing jointly
    J a head of household

9 Last year Ron, a single man, made $22,000. This year he made $25,000. According to these tables, how much more will he pay in taxes this year?

    A $450
    B $3,300
    C $3,700
    D $840

Exponents, Powers, and Roots

Products and Factors

You can write 10×10 as 10^2 "ten squared" or "ten to the second power"

$8 \times 8 \times 8$ as 8^3 "eight cubed" or "eight to the third power"

$5 \times 5 \times 5 \times 5$ as 5^4 "five to the fourth power"

6 as 6^1 "six to the first power"

Each of the numbers 2, 3, 4, and 1 is called an **exponent.** An exponent tells how many times a factor appears in a product. In an expression like 10^2 or 10^3, the number 10 is called the **base** and the entire expression is called a **power.**

Examples:

$$8^2 = 8 \times 8 = 64 \qquad\qquad (-2)^4 = (-2) \times (-2) \times (-2) \times (-2) = 16$$

$$\left(\frac{1}{3}\right)^5 = \frac{1}{3} \times \frac{1}{3} \times \frac{1}{3} \times \frac{1}{3} \times \frac{1}{3} = \frac{1}{243} \qquad -2^4 = -(2^4) = -(2 \times 2 \times 2 \times 2) = -16$$

Finally, the value of any nonzero number with an exponent of zero is the number 1.

Examples: $5^0 = 1$, $25^0 = 1$, $\left(\frac{1}{5}\right)^0 = 1$, $1^0 = 1$

PRACTICE

Use exponents to write each product.

1    $4 \times 4 \times 4$

2   3×3

3    $5 \times 5 \times 5 \times 5$

4    $2 \times 2 \times 2 \times 2 \times 2$

5   ten cubed

6   three to the fourth power

7    $-12 \times (-12) \times (-12)$

8   0.8×0.8

Find the value of each power.

9   2^3

10    $(-4)^2$

11   three to the fourth power

12   10^5

13   five squared

14   eleven to the third power

Fill in each blank below.

15   What is the exponent in 4^9? _____

16   Which is greater, 2^3 or 2^5? _____

17   Complete the number sentence: $\left(\frac{1}{2}\right)^5 = \underline{\;?\;}.$ _____

18    $9^0 = \underline{\;?\;}$ _____

19    $123^1 = \underline{\;?\;}$ _____

20   In words, the expression 12^6 means $\underline{\;?\;}$.

_____

_____

Solving Problems with Powers

Always evaluate exponents before you add, subtract, multiply, or divide.

$$9^2 + 2 = 81 + 2 = 83$$

$$3^3 - 4^2 = 27 - 16 = 11$$

$$(-2)^2 \times 3 = 4 \times 3 = 12$$

$$6^2 \div 2^2 = 36 \div 4 = 9$$

PRACTICE

Solve each problem below. When a section of a problem appears in parentheses, use that entire expression as the base. Example: $(2 + 3)^2 = (5)^2 = 25$

1 $2^3 + 2 =$ _____

2 $3 + 4^2 =$ _____

3 $(-3)^2 + 4^3 =$ _____

4 $5^3 - 5 =$ _____

5 $(-7)^2 - 2^2 =$ _____

6 $6^3 - 6^2 =$ _____

7 $(30 - 25)^3 =$ _____

8 $(4 - 2)^3 =$ _____

9 $\left(\frac{1}{8}\right)^2 \times 2^2 =$ _____

10 $10^2 \div 2 =$ _____

11 $9^2 \div 3 =$ _____

12 $(-2)^4 \div 4^2 =$ _____

13 $7^2 \div 19^0 =$ _____

14 $6^1 \times 3^2 =$ _____

15 $5 + 6^2 + 2^3 =$ _____

16 $(-1)^2 + 5^2 + 10^2 =$ _____

17 $9^1 + 3^2 + 4^2 =$ _____

18 $(8 + 2)^3 =$ _____

19 $\left(\frac{1}{5}\right)^2 \times 5 =$ _____

20 $\left(\frac{1}{6}\right)^2 \times 3^1 =$ _____

21 $\left(\frac{1}{3} - \frac{1}{9}\right)^2 =$ _____

22 $(7 + 3)^2 + 5 =$ _____

Simplifying Powers

If two powers have the same base, such as 3^8 and 3^2, you can simplify their product or their quotient.

When powers with the same base are multiplied, add the exponents.

$$3^8 \times 3^2 = (3 \times 3 \times 3 \times 3 \times 3 \times 3 \times 3 \times 3) \times (3 \times 3)$$
$$= 3 \times 3 \times 3 \times 3 \times 3 \times 3 \times 3 \times 3 \times 3 \times 3$$
$$= 3^{10}$$

This rule works for powers with an exponent of 1:
$$5 \times 5^4 = 5^1 \times 5^4 = (5) \times (5 \times 5 \times 5 \times 5) = 5^5$$

When powers with the same base are divided, subtract the exponents.

$$\frac{3^8}{3^2} = \frac{3 \times 3 \times 3 \times 3 \times 3 \times 3 \times 3 \times 3}{3 \times 3}$$

$$= \frac{3 \times 3}{3 \times 3} \times \frac{3 \times 3 \times 3 \times 3 \times 3 \times 3}{1}$$

$$= 3^6$$

$$\frac{3^2}{3^8} = \frac{3 \times 3}{3 \times 3 \times 3 \times 3 \times 3 \times 3 \times 3 \times 3}$$

$$= \frac{3 \times 3}{3 \times 3} \times \frac{1}{3 \times 3 \times 3 \times 3 \times 3 \times 3}$$

$$= \frac{1}{3^6}$$

PRACTICE

Simplify each expression below, writing your answers with exponents. If a problem cannot be simplified, write "Cannot be simplified."

Sample: $2^6 \div 2^4 = 2^2$

1 $6^3 \times 6^2 = $ _____

2 $5^4 \div 5^2 = $ _____

3 $16^8 \div 16^3 = $ _____

4 $50^3 + 50^2 = $ _____

5 $12^3 \times 12 = $ _____

6 $7^6 \times 12^6 = $ _____

7 $14^3 \div 12^3 = $ _____

8 $\dfrac{52^3}{52^6} = $ _____

9 $\dfrac{6^{10}}{6^3} = $ _____

10 $10^3 \times 10^5 \times 10 = $ _____

11 $67^{12} \div 67^3 = $ _____

12 $71 \times 71 \times 71^4 = $ _____

13 $\dfrac{14^{17}}{14^{32}} = $ _____

14 $\dfrac{8^8}{8^4} = $ _____

Scientific Notation

When you multiply a number by a power of ten, a shortcut is to move the decimal point one place to the right for every zero.

$$2.34 \times 10 = 23.4$$
$$2.34 \times 1,000 = 2,340$$
$$2.34 \times 1,000,000 = 2,340,000$$

If you write the same examples using exponents, you can see that the exponent tells you how many places to move the decimal point.

$$2.34 \times 10^1 = 23.4$$
$$2.34 \times 10^3 = 2,340$$
$$2.34 \times 10^6 = 2,340,000$$

Scientific notation is a method of writing numbers using powers of ten.

$$3,400,000,000 \quad \text{is written} \quad 3.4 \times 10^9$$
$$50,600 \quad \text{is written} \quad 5.06 \times 10^4$$

You also can write decimals less than 1 in scientific notation:
$$6.0 \times 10^{-3} \quad \text{means} \quad 0.006$$
$$1.05 \times 10^{-9} \quad \text{means} \quad 0.00000000105$$

The negative sign in the exponent means that the decimal point is moved to the left.

PRACTICE

Rewrite each number so it *is not* in scientific notation.

1 3.05×10^6 _____

2 4.1×10^{11} _____

3 2.0035×10^{10} _____

4 4.5×10^{-5} _____

5 9.105×10^5 _____

6 7.31×10^{-2} _____

7 8.0×10^{-9} _____

8 9.0×10^4 _____

Rewrite each number so it is written in scientific notation.

9   5,600,000,000   _____

10   3,400,000   _____

11   505,000,000,000   _____

12   0.0000000192   _____

13   60,710,000,000,000   _____

14   0.00000004006   _____

15   0.000000000000051698   _____

Circle the larger number in each pair below.

16 1.9×10^6 1.9×10^9

17 5.4×10^6 6.7×10^3

18 1.002×10^{-12} 1.002×10^1

19 8.7×10^{-13} 9.014×10^{-20}

20 9.1×10^{-6} 8.1×10^{-6}

21 3.0012×10^6 3.012×10^6

22 8.1117×10^{-12} 5.7×10^{-14}

23 9.9×10^{-6} 9.9×10^{-16}

24 7.0102×10^8 8.45×10^{-5}

Rewrite each number so it *is not* in scientific notation. Then add, subtract, multiply, or divide as indicated.

25 $(1.05 \times 10^{-3}) + (1.4 \times 10^{-2}) =$ _____

26 $(6.7 \times 10^3) - 500 =$ _____

27 $\dfrac{7.09 \times 10^2}{10} =$ _____

28   One drop of a liquid weighs just 5.0×10^{-2} grams. How many drops are in 1 gram of the liquid?

29   A type of bead weighs 6.1×10^{-3} ounces. A shopkeeper wants to put the beads into packages of about 100. How many ounces of beads should she put in each bag?

30   City officials estimate that about 1.2×10^4 people came to the fireworks display last year. This year it rained, and only 8×10^3 people came. How much larger was last year's crowd?

Square Root

Since $2 \times 2 = 4$, the number 2 is called the "square root" of 4.
Similarly, $3 \times 3 = 9$, so the square root of 9 (or $\sqrt{9}$) is 3.

$10 \times 10 = 100$, so $\sqrt{100} = 10$.

$\frac{1}{3} \times \frac{1}{3} = \frac{1}{9}$, so $\sqrt{\frac{1}{9}} = \frac{1}{3}$.

$0.5 \times 0.5 = 0.25$, so $\sqrt{0.25} = 0.5$.

Perfect Squares and their Square Roots

A "perfect square" is a number whose square root is a whole number. Here are the first sixteen perfect squares and their square roots.

$\sqrt{1} = 1$	$\sqrt{25} = 5$	$\sqrt{81} = 9$	$\sqrt{169} = 13$
$\sqrt{4} = 2$	$\sqrt{36} = 6$	$\sqrt{100} = 10$	$\sqrt{196} = 14$
$\sqrt{9} = 3$	$\sqrt{49} = 7$	$\sqrt{121} = 11$	$\sqrt{225} = 15$
$\sqrt{16} = 4$	$\sqrt{64} = 8$	$\sqrt{144} = 12$	$\sqrt{256} = 16$

You can use perfect squares to estimate the square root of other numbers. For example, here is how to estimate $\sqrt{10}$.

1. Find a perfect square less than 10 and a perfect square greater than 10: 10 is between 9 and 16.
2. You can conclude that $\sqrt{10}$ is between $\sqrt{9}$ and $\sqrt{16}$.
3. That means $\sqrt{10}$ is between 3 and 4.

PRACTICE

1 $\sqrt{9} = $ _____

2 $\sqrt{16} = $ _____

3 $\sqrt{121} = $ _____

4 $\sqrt{36} = $ _____

5 $\sqrt{100} = $ _____

6 $\sqrt{225} = $ _____

7 $\sqrt{64} = $ _____

8 $\sqrt{196} = $ _____

9 Between what two whole numbers is $\sqrt{7}$?

10 Between what two whole numbers is $\sqrt{12}$?

11 Between what two whole numbers is $\sqrt{19}$?

12 Which of these is closest to $\sqrt{75}$?

A 4   C 9
B 6   D 10

13 Which of these is closest to $\sqrt{105}$?

F 6   H 10
G 8   J 12

To solve each problem below, first find the square roots. Then add, subtract, multiply, or divide as indicated.

14 $\sqrt{4} + \sqrt{9} = $ _____

15 $\sqrt{25} - \sqrt{16} = $ _____

16 $\sqrt{4} \times \sqrt{16} = $ _____

17 $\dfrac{\sqrt{36}}{\sqrt{100}} = $ _____

Exponents, Powers, and Roots Skills Practice

Circle the letter for the best answer to each problem.

1

$10^8 \div 10^4 = $ _____

A 10^2
B 10^4
C 1^2
D 1^4
E None of these

2

$14 \times 14 \times 14 = $ ___

F 14^3
G 42^3
H 3^{14}
J 3×14
K None of these

3

$5^3 \times 10^2 = $ _____

A 1,500
B 1.25
C 150
D 12,500
E None of these

4

$(-2)^4 + 4^2 = $ _____

F 16
G −8
H 8
J 32
K None of these

5

$7^2 \times 7^4 = $ _____

A 7^2
B 7^3
C 7^6
D 1^6
E None of these

6

$(1 + 4)^2 = $ _____

F 9
G 17
H 25
J 10
K None of these

7

$5.105 \times 10^2 = $ ___

A 51.05
B 510.5
C 0.05105
D 5.105
E None of these

8

$1,000 \times 7.124 = $ ___

F 7,124,000
G 7,124
H 0.007124
J 7.124
K None of these

9 Which number has a square root that is a whole number?

A 15
B 20
C 36
D 50

10 Which of these is another way to write 80,100,000?

F 801×10^5
G 8.01×10^5
H 8.01×10^7
J 80.1×10^7

11 A box of 100 straight pins weighs 3 ounces. How much does each pin weigh?

A 3×10^{-2} oz
B 3×10^2 oz
C 3×10^{-1} oz
D 3×10^{-3} oz

12 In the year 1650, there were 5.5×10^8 people living on Earth. In 1997, the Earth's population was 5.9×10^9 people. How much had the population grown?

F 5.35×10^9
G 4×10^9
H 4×10^8
J 5.35×10^8

13 Which of these number sentences is true?

    A $9.5 \times 10^5 < 4.5 \times 10^5$
    B $9.5 \times 10^5 < 9.5 \times 10^6$
    C $9.5 \times 10^5 < 9.5 \times 10^{-8}$
    D $9.5 \times 10^5 < 9.5 \times 10^4$

14 Which of these has the same value as 134^1?

    F 134   **H** $\frac{1}{134}$
    G 0   **J** 1

15 Donica needs 2.5 square yards of fabric to make a table cloth. She needs another 7 square yards for a bedspread. How many square yards of fabric does she need altogether?

    A 55.25   **C** 9.5^4
    B 9.5   **D** 90.25

16 Kim just started his own company. He has 2 employees, and he hopes to double the size of the company every year. If that happens, he will have 2^2 employees in his second year, 2^3 in his third year, 2^4 in his fourth year, and so on. If the company grows at this rate, how many employees will Kim have in his sixth year?

    F 12   **H** 32
    G 36   **J** 64

17 Which of these fractions has the same value as $\frac{10^8}{10^{10}}$?

    A $\frac{1}{2}$ **C** $\frac{1}{100}$
    B $\frac{4}{5}$ **D** $\frac{1}{10}$

18 Which of these is another way to write 8×10^{-4}?

    F −80,000   **H** 0.0008
    G 0.0048   **J** −0.0008

19 Which of the following has a negative value?

    A $(-2)^{22}$ **C** 5^{-3}
    B $(-2)^9$ **D** 5^{-4}

20 Which of the following has the same value as 147×10^0?

    F 147   **H** 0
    G 1   **J** 10

21 What is the sum of 10 squared and 5 cubed?

    A 225   **C** 125
    B 2,500   **D** 250

22 The square root of 90 lies between what two whole numbers?

    F 9 and 10
    G 10 and 11
    H 8 and 9
    J 11 and 12

23 Which of these has the same value as $\sqrt{36} + \sqrt{100}$?

    A 1,290
    B 63
    C 31
    D 16

Algebra

Patterns

A key to understanding math is to see and use patterns. When you describe a pattern using words and symbols, you are using the math ideas called **algebra.**

PRACTICE

1 How many bricks are used in the next figure in the sequence below?

_____

2 Draw the next figure in the sequence above.

3 The missing figure in the sequence below would have _?_ columns, and would have _?_ bricks in each column.

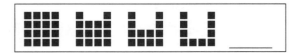

_____, _____

4 Circle the letter for the choice that describes the pattern in the box below.

A thick, thin, thick, thin
B thick, thick, thin, thick, thick, thin
C thick, thin, thin, thick, thin, thin

5 If you start with 5 and multiply each number by 3, you create this list:
    15, 45, 135, 405, 1215, ...
Which of the following numbers *does not* belong in this sequence?

F 10,935
G 32,805
H 98,410

6 Find the next number in this sequence:
    1, 2, 4, 7, 11, 16, ...

_____

7 In the following pattern, each new number is the last number times 2:
    1, 2, 4, 8, 16, 32, ...
Which number will eventually appear in the list above?

A 55   **C** 99
B 63   **D** 128

8 For the following sequence, first find the missing number. Then tell how to get each number from the previous number.
    2700, 900, _?_, 100, $33\frac{1}{3}$

_____

9 What number is missing from this sequence?
    0, 10, 30, 60, 100, _____, 210, ...

_____

Completing Number Sentences

There are two **properties of one.**

First, if you multiply or divide a number by one, you do not change the value.
$$15 \times 1 = 15 \qquad 20 \div 1 = 20$$

Second, if you divide any (nonzero) number by itself, the quotient is one.
$$15 \div 15 = 1$$

There are several **properties of zero.**

If you add or subtract zero from a number, you do not change the value.
$$15 + 0 = 15 \qquad 20 - 0 = 20$$

If you subtract a number from itself, the difference is zero. $15 - 15 = 0$

If you multiply any number by zero, the product is zero. $20 \times 0 = 0$

If you divide zero by any (nonzero) number, the quotient is zero. $0 \div 15 = 0$

An expression like "$15 \div 0$" has no meaning.

PRACTICE

Write +, −, ÷, or × in each box below to complete the number sentences. Always work inside parentheses before you work outside the parentheses.

Whole Numbers (If you have difficulty with the problems below, review pages 1–19.)

1 $100 \;\square\; 10 = 10$

2 $67 \;\square\; 67 = 0$

3 $13 \;\square\; 3 = 69$

4 $(3 + 3) \;\square\; 2 = 12$

5 $2 \times (3 \;\square\; 3) = 18$

6 $105 \;\square\; 25 = 80$

7 $920 \;\square\; 0 = 920$

8 $4 \times (568 \;\square\; 568) = 4$

9 $3 \times (17 \;\square\; 14) = 9$

10 $40 + (3 \;\square\; 3) = 46$

Decimals (For review, see pages 20–31.)

11 $1.9 \;\square\; 100 = 190$

12 $0.4124 \;\square\; 2.1 = 2.5124$

13 $20 \;\square\; 0.5 = 10$

14 $20 \;\square\; 0.5 = 40$

Fractions (For review, see pages 32–46.)

15 $\frac{1}{5} \;\square\; \frac{1}{5} = \frac{1}{25}$

16 $\frac{3}{10} \;\square\; \frac{1}{10} = \frac{2}{5}$

17 $12 \;\square\; \frac{1}{3} = 11\frac{2}{3}$

18 $5 \;\square\; \frac{1}{5} = 25$

Signed Numbers (For review, see pages 47–53.)

19 $-12 \;\square\; 8 = -20$

20 $4 \;\square\; -9 = -5$

21 $-2 \;\square\; -3 = -5$

22 $-10 \;\square\; -10 = 1$

23 $-4 \;\square\; -4 = 0$

Exponents, Powers, and Roots (For review, see pages 78–85.)

24 $3.102 \;\square\; 10^5 = 310{,}200$

25 $560 \;\square\; 10^2 = 5.6$

26 $7.19 \;\square\; 10^{-3} = 0.00719$

27 $10^2 \;\square\; 10^3 = 10^5$

Inverse Operations

Addition and subtraction are **inverse operations.** Each one will "undo" the other.

$$12 + 30 = 42$$
$$\text{so } 42 - 30 = 12$$
$$\text{and } 42 - 12 = 30$$
$$17 - 13 = 4$$
$$\text{so } 4 + 13 = 17$$

Multiplication and division are inverse operations for each other.

$$4 \times 5 = 20$$
$$\text{so } 20 \div 5 = 4$$
$$\text{and } 20 \div 4 = 5$$
$$60 \div 3 = 20$$
$$\text{so } 20 \times 3 = 60$$

To solve a problem that has a missing number, you may be able to use the inverse of the given operation.

Problem: $145 + \square = 235$

Solution:

Use the inverse: If $145 + \square = 235$, then $235 - 145 = \square$.

Now you can solve the problem:

$$\begin{array}{r} 235 \\ -\ 145 \\ \hline 90 \end{array}$$

To check: $145 + \boxed{90} = 235$. It checks.

PRACTICE

For each problem, write a number in the box that makes the number sentence true. Try to use inverse operations to solve the problems.

1 $105 + \square = 199$

> *Hint:* What is $199 - 105$?

2 $311 + \square = 527$

3 $89 + \square = 113$

4 $\frac{1}{5} + \square = \frac{9}{25}$

5 $0.33 + \square = 1.6$

6 $-12 + \square = -40$

7 $0 + \square = -17$

8 $\square - 16 = 42$

> *Hint:* What is $42 + 16$?

9 $\square - 59 = 96$

10 $\square - 12 = 120$

11 $\square - \frac{1}{2} = 1\frac{1}{2}$

12 $\square - \frac{1}{4} = \frac{1}{3}$

13 $\square - 2.01 = 9.99$

14 $32 - \square = 17$

> *Hint:* What is $32 - 17$?

15 $66 - \square = 59$

16 $2.1 - \square = 0.08$

17 $4\frac{4}{5} - \square = \frac{1}{10}$

18 $22 \times \square = 132$

> *Hint:* What is $132 \div 22$?

19 $3.3 \times \square = 0.099$

20 $4 \times 3 \times \square = 60$

> *Hint:* What is $60 \div 12$?

21 $2 \times 10 \times \square = 140$

22 $96 \times \square = 16$

> *Hint:* What is $16 \div 96$?

23 $\square \times 15 = 180$

Writing Letters and Symbols for Words

An algebra problem uses letters to take the place of numbers. These letters are called **variables** or **unknowns.** A letter can represent a single number or it can represent many values.

In	3	4	5	7
Out	1	2	3	5

In this table, suppose that x represents each "In" number. Then each "Out" number can be described as $x - 2$.

Example:

a number increased by five

If you use the letter n to represent "a number," then this phrase can be described as $n + 5$.

When variables and numbers are put together with an operation like addition, they create an **algebraic expression.**

- Addition and subtraction are shown with the usual symbols [+ and −].
- Multiplication is usually shown with no sign at all [$4x$ instead of $4 \times x$] or with parentheses [the expression $3(x + y)$ represents 3 times the sum of x and y].
- Division is usually shown in fraction form [$\frac{n}{5}$ represents $n \div 5$].

Algebraic Expressions	
Expression	**Meaning**
$n + 2$	two more than n
$a - 6$	six less than a
$6 - a$	a fewer than six
$8x$	eight times x
$\frac{7}{b}$	seven divided by b
$\frac{b}{7}$	b divided by 7
$\frac{3c}{10}$	three-tenths of c or one-tenth of $3c$

PRACTICE

Write an algebraic expression for each phrase below. Use the variable x to stand for the unknown.

1   a number divided by fourteen _____

2   a number increased by two _____

3   fifty-four times a number _____

4   a number squared _____

5   sixty-three less than a number _____

6   the sum of a number and 42 _____

7   half of a number _____

8   the product of a number and twelve _____

9   a number doubled _____

10   the difference between a number and one hundred _____

11   negative two minus a number _____

12   a number divided by thirteen _____

13   negative four divided by a number _____

14   one over the square root of a number _____

15   two-fifths of a number _____

Writing Longer Algebraic Expressions

A long word problem often can be summed up with just a few algebraic symbols.

Problem 1:

A number is divided by 4 and then 16 is added. The result is 13. What is the original number?

Algebraic Expression

Use n for the original number.
$$\frac{n}{4} + 16 = 13$$

Problem 2:

Angela is shopping. She begins with $30.00 in her purse, and she must save $1.50 to pay for her train ride home. How much money does she have left to spend after paying $5.50 to see a movie?

Use x for the amount she has left to spend.
$$30 - 5.50 - x = 1.50$$

When you write a complex algebraic expression, you can use parentheses to show what operation should be done first. For instance, $3x + 2$ means "Multiply 3 and x, and then add 2." That is different from $3(x + 2)$ which means "Add x and 2, and multiply the sum by 3."

Another way to see the difference between $3x + 2$ and $3(x + 2)$ is to substitute 4 for x in each expression:

$$3x + 2 = 3(4) + 2 = 12 + 2 = 14$$
$$3(x + 2) = 3(4 + 2) = 3 \times 6 = 18$$

When there are no parentheses, multiplication and division are done before addition and subtraction.

PRACTICE

Circle the letter for the correct algebraic expression for each sentence below.

1 Add ten to the product of five times n.

  A $5n + 10$
  B $5(n + 10)$
  C $5n + 10n$

2 Divide twelve by a, then add two.

  F $\frac{a}{12} + 2$
  G $\frac{12}{a} + 2$
  H $\frac{12}{(a+2)}$

3 Subtract seven from three times b. Then add forty.

  A $(3b - 7) + 40$
  B $(7 - 3b) + 40$
  C $7 - 3(b + 40)$

4 Add fifteen to two-thirds of x.

  F $15 + \frac{2}{3} + x$
  G $x\left(15 + \frac{2}{3}\right)$
  H $\frac{2x}{3} + 15$

5 Multiply negative 12 by the sum of n plus five.

  A $-(12n) + 5$
  B $-12(n + 5)$
  C $-12n + 5$

6 Square the difference between twenty-four and b.

  F $(24 - b)^2$
  G $24^2 - b$
  H $24 - b^2$

Circle the letter for the sentence that represents each algebraic expression below.

7 $\dfrac{5}{x^2} + 7$

   A Divide five by the square of x. Then add seven.
   B Add seven to the square of x. Then divide five by the sum.
   C Divide five by x and square the quotient. Then add seven.

8 $\dfrac{c - 5}{90}$

   F Subtract $\dfrac{5}{90}$ from c.
   G Divide c by ninety. Then subtract five.
   H Subtract five from c. Then divide the difference by 90.

9 $n(35 - n)$

   A Multiply n by 35. Then subtract n.
   B Subtract n from 35. Then multiply the difference by n.
   C Subtract $35n$ from n.

10 $x^3 + (3 - 5)$

   F Subtract 3 from the cube of x. Then subtract 5.
   G Cube x. Then add –2.
   H Cube x. Then subtract –2.

Write your own algebraic expression for each sentence below.

11 Add twice n to four. _____

12 Subtract five from the square root of n. _____

13 Add thirty to one-half n. _____

14 Multiply four times the quantity n squared minus two. *Hint:* The words "the quantity" are signal words for parentheses. The phrase "the quantity n squared minus two" represents $(n^2 - 2)$. _____

15 Add n to four, multiply the sum by three, and then subtract twelve. _____

16 Divide the expression n minus five by n. _____

17 Multiply negative $\dfrac{1}{3}$ times the quantity n plus five. _____

18 Add eight to one fifth of n. _____

19 Subtract three times n from sixty. _____

20 Subtract ten from the product of seven times n. _____

21 Subtract twelve from the square root of a number. _____

22 Divide the quantity six less than n by twelve. _____

Writing Longer Algebraic Expressions

Writing Algebraic Equations

If you use an algebraic equation to describe a situation in a word problem, you can solve the word problem.

Problem: Each member of the sales staff at the Daily News makes $4.00 per hour plus a $0.75 commission for every newspaper subscription sold. Last week, Len worked 32 hours and made $256.50. How many subscriptions did he sell?

Solution:
We know that wages + commission = $265.50.
Let s represent the number of subscriptions he sold. Then:
  wages = wage per hour × number of hours
  commission = price per subscription × number of subscriptions
An equation is
$$(4)(32) + (0.75)(s) = \$256.50$$
or $\qquad 128 + 0.75s = 256.50$

PRACTICE

Write an algebraic equation to describe each situation below.

1 Cori makes twice as much money as her brother. Their combined earnings are $72,000 per year. (Let b stand for the brother's earnings.)

2 When Tom's car was repaired, it cost $675. The garage charged $159 for materials and $45 per hour for labor. (Let h stand for the number of hours they worked on the car.)

3 Elena's company gives employees one dollar in company stock for every two dollars in stock that they buy themselves. In this way, Elena has acquired $450 in company stock. (Let m stand for the amount of her own money Elena has spent on stock.)

4 If Anne averages 75 miles an hour for 4 hours of driving and 65 miles an hour for 2 hours of driving, how far will she get? (Let d stand for the total distance she travels.)

5 Leo pays a basic telephone fee of $25.00 a month. He must pay $0.04 for every local call he makes over 40 in a month. What equation can he use to figure his local phone bill each month? (Let b stand for his total bill. Let c stand for the number of local calls he makes.)

6 The policy at Royko Sprockets is to pay the company president 25 times the salary of the lowest-paid employee. Currently, the president makes $300,000 dollars more than the lowest-paid employee. (Let L stand for the salary of the lowest-paid employee.)

7 Yukio bikes one morning. She travels 9 miles an hour. Her husband sets out an hour later and averages 15 miles an hour. Find t, the number of hours it takes Yukio's husband to catch up with her. (*Hint:* At that point they will have traveled the same distance.)

Solving an Equation with an Addition or Subtraction Sign

To find the solution to an algebraic equation, you must get the variable alone on one side of the equation.

Equation:	$x - 37 = 76$
Add 37 to each side:	$\quad 37 \quad\; 37$
The solution is $x = 113$.	$x \qquad = 113$

Notice that the same number was added to both sides of the equation. An equation works like a balance: If you do the same thing to both sides, the two sides stay in balance and remain equal.

To solve an equation, you work with inverse operations.

To solve an addition problem, you subtract.

Equation: $x + 3 = 15$

Subtract 3 from each side. $\quad -3 \quad\; -3$

$$x + 0 \qquad 12$$

Adding 0 to a number does not change its value, so $\quad x \quad = 12$

To solve a subtraction problem, you add.

$$x - 3 = 6$$
$$+3 \quad +3$$
$$x \qquad = 9$$

Some problems need two steps.

$$15 - x = 7$$
$$+x \qquad\quad +x$$
$$15 \quad = 7 + x$$
$$-7 \qquad\quad -7$$
$$8 \quad = \quad x$$

> ***Remember:*** To solve an equation, you must get the variable alone on one side of the equation.
>
> To add two numbers that have the same sign, add their absolute values and give the sum their common sign.
>
> To add a positive number and a negative number, subtract their absolute values and give the sum the sign of the number whose absolute value is larger.
>
> To subtract signed numbers, change the sign of the number being subtracted. Then add the two signed numbers.

PRACTICE

Find the value of x in each equation below.

1 $x + 95 = 270$

8 $42 - x = 23$

2 $x + 113 = 605$

9 $14 - x = -2$

3 $x - 93 = 25$

10 $82 - x = 0$

4 $17 = x + 9$

11 $7.81 - x = 2.95$

5 $39 - x = 0$

12 $6 + x - 5 = 30$

6 $16 + x = 12$

13 $19 - x - 5 = 2$

7 $x - 2.5 = 10.2$

14 $x + 15 + 32 = 70$

For the problems below, you add or subtract groups of letters, numbers, or powers. These groups are called *terms*. Be sure that you add or subtract the same group on each side of the equation.

Sample:

$$
\begin{aligned}
x + 2x &= 2x + 3 \\
-2x \quad &\quad -2x \\
\hline
x + 0 &= 0 + 3 \\
x &= 3
\end{aligned}
$$

15 $x + \dfrac{x}{3} = 5 + \dfrac{x}{3}$

16 $14 + 3x = x + 3x$

17 $x + b + 3x^2b = b + 49 + 3x^2b$

18 $32 + (x - x) = (x - x) + x$

19 $14x + 12b^2 - x + 3 = 14x + 12b^2$

Solving an Equation with an Addition or Subtraction Sign

Solving an Equation with a Multiplication or Division Sign

To solve a multiplication equation, you can divide.	To solve a division equation, you can multiply.	To check, replace the variable with your solution. Then work *separately* on each side of the equation.
Problem: $42x = 84$	**Problem:** $\frac{x}{5} = 13$	To check $x = 65$ in $\frac{x}{5} = 13$, substitute 65 for x.
Divide by 42: $\frac{42x}{42} = \frac{84}{42}$	Multiply by 5: $(5)\left(\frac{x}{5}\right) = (5)(13)$	Does $\frac{65}{5} = 13$?
$x = 2$	$x = 65$	Does $13 = 13$? Yes. The solution $x = 65$ checks.

PRACTICE

Solve each equation below. Then replace the variable in the original equation to check your work.

1 $23n = 46$

2 $105 = 5x$

3 $-16a = 80$

4 $\frac{n}{6} = 36$

5 $\frac{c}{12} = 12$

6 $\frac{40}{a} = 5$

7 $\frac{15.3}{n} = 3$

8 $\frac{x}{14} = 5$

9 $\frac{p}{30} = 25$

10 $\left(\frac{7}{3}\right)x = 2$

Each of the next equations has two steps. To solve, first "undo" the addition or subtraction. Then "undo" the multiplication or division.

Sample:
$$2x + 5 = 15$$
$$\underline{\quad - 5 \quad - 5}$$
$$2x \quad\quad = 10$$
$$\frac{2x}{2} = \frac{10}{2}$$
$$x = 5$$

11 $4x - 12 = 8$

12 $10n + 12 = 112$

13 $\frac{b}{15} + 6 = 8$

14 $\frac{t}{2} - 10 = 6$

15 $12c - 14 = 106$

16 $\left(\frac{2}{3}\right)s + 2 = 10$

Reminder: Dividing by the fraction $\frac{2}{3}$ is the same as multiplying by fraction $\frac{3}{2}$. The two fractions $\frac{2}{3}$ and $\frac{3}{2}$ are called **reciprocals.**

17 $\left(\frac{4}{5}\right)v - 7 = 9$

18 $\frac{3x}{4} + 3 = 15$

Combining Terms

A number grouped with one or more variables is called a **term.** Each of the expressions $2z$, $5ab$, and $6x^2y$ are all terms. A term does not contain a $+$ sign, a $-$ sign, or an $=$ sign.

If two terms have the same variables and exponents, such as $3x^2y$ and $12x^2y$, they are called **like terms.** You can add or subtract like terms.

$$2a + 3a = 5a$$
$$3x^2 - 2x^2 = x^2$$
$$3cb^2 + 7cb^2 + 2cb^2 = 12cb^2$$

Terms such as $6ab$ and $6b$, or terms such as $12a^2b$ and $6ab^2$, are called **unlike terms.** You cannot add or subtract unlike terms.

$6ab + 6b = \underline{\ ?\ }$ You cannot combine the two terms.
$5n^2 - n = \underline{\ ?\ }$ You cannot combine the two terms.

You always can multiply or divide terms whether they are like terms or unlike terms.

$$2ab \times 3ab^2 = (2)(3)(a)(a)(b)(b^2) = 6a^2b^3$$

$$\frac{10x^2yz^5}{2xz^2} = \frac{10}{2} \times \frac{x^2}{x} \times \frac{y}{1} \times \frac{z^5}{z^2} = 5xyz^3$$

PRACTICE

Combine these terms as indicated. Write "Cannot be combined" if the terms cannot be combined.

1 $5ab + 2ab =$ __

2 $3a + 4b =$ ____

3 $4x(3x) =$ _____

4 $-2a(3a) =$ ____

5 $6x - 2x^2 =$ ____

6 $-4n(-4a) =$ ____

7 $5b^2(2b^5) =$ ___

8 $\frac{2a}{a} =$ ____

9 $\frac{x^2}{x} =$ _____

10 $\frac{4t^2s}{ts} =$ _____

11 $3ab + 5b =$ ___

12 $\frac{4x^3y^2}{4x^2y} =$ _____

13 $3ab(a^2) =$ _____

14 $\frac{16u^3q^2}{2u^2q^2} =$ ____

15 $\frac{8xy^2}{2x} =$ _____

Solving an Equation with Parentheses

A common pattern in algebra is to multiply a number or variables times a sum or difference. To remove the parentheses, you multiply that number or variable by each term inside the parentheses. The pattern looks like the following:

$$a(b + c) = ab + ac$$
$$a(b - c) = ab - ac$$

Below, Examples 1 and 2 are simple illustrations of the pattern. Examples 3 and 4 are more complex illustrations of the same pattern.

Example 1:

$$3(x - 5) = 20$$
$$3(x) - 3(5) = 20$$
$$3x - 15 = 20$$

Example 2:

$$5(3a + 6) = 40$$
$$5(3a) + 5(6) = 40$$
$$15a + 30 = 40$$

Example 3:

$$12a(a + b) = 124$$
$$12a(a) + 12a(b) = 124$$
$$12a^2 + 12ab = 124$$

Example 4:

$$3ab(a - b^3) = 230$$
$$3ab(a) - 3ab(b^3) = 230$$
$$3a^2b - 3ab^4 = 230$$

PRACTICE

Eliminate the parentheses in each problem. Then solve the problem.

1 $5(x + 2) = 510$

2 $3(x - 5) = 30$

3 $10 = 5(b - 3)$

4 $28 = 2(7 + n)$

5 $\frac{1}{3}(v - 2) = \frac{1}{3}$

6 $12 = \frac{1}{5}(y + 10)$

7 $4(2c + 5) = 28$

8 $\frac{1}{8}(8d - 16) = 98$

9 $7(3 + n) = 140 + 21$

10 $1.5(2x + 5) = 16.5$

11 $-2(e - 4) = -4$

12 $4(-f + (-5)) = 0$

13 $3a\left(\frac{1}{a} + 5\right) = 33$

14 $\frac{1}{b}(2b - 10) = 17$

15 $-5(-10 - a) = 60$

For problems 16–21, multiply to eliminate the parentheses. Do not try to solve the problems.

16 $3x(x + 5)$

17 $2ab(a + b)$

18 $4c(11 - c^2)$

19 $y(2xy + y^3)$

20 $2ab(2a + 2b)$

21 $3x^2(2x + 2x^2)$

When a Sum or Difference Appears in a Fraction

If a sum or a difference appears in a fraction, divide each term separately. This will not change the value of the expression.

Then you can simplify each term.

$$\frac{4 + 6x}{2} = 42$$

$$\frac{4}{2} + \frac{6x}{2} = 42$$

$$2 + 3x = 42$$

$$3x = 40$$

$$x = \frac{40}{3} = 13\frac{1}{3}$$

Remember, any nonzero number divided by itself is 1, and any value multiplied by 1 is that same value.

$$\frac{2ab}{a} = 2b \text{ and } \frac{a}{a} = 1$$

$$\frac{2ab + a}{a} = 5$$

$$\frac{2ab}{a} + \frac{a}{a} = 5$$

$$2b + 1 = 5$$

$$2b = 4$$

$$b = 2$$

PRACTICE

Divide to simplify each equation below. Then solve the equation.

1. $\dfrac{10 + 5x}{5} = 12$

2. $\dfrac{4 + 4x}{4} = 9$

3. $\dfrac{10 + 2x}{2} + 5 = 20$

4. $\dfrac{2(x + 4)}{4} = 5$

5. $\dfrac{2x - 6}{12} = 0$

6. $\dfrac{5x + 10}{5} = 1$

Divide to simplify each expression below. Do not try to solve.

7. $\dfrac{3x - x^2}{x}$

8. $\dfrac{5xy + y}{xy}$

9. $\dfrac{2xy - x^2y}{xy}$

10. $\dfrac{2x - 4x^2}{2x}$

11. $\dfrac{x + x^2}{x} + 14$

12. $\dfrac{3x^3 + 9x^2}{x^2}$

When a Variable Appears Twice in an Equation

If a variable or unknown appears more than once in an equation, then the equation has like terms. You must always combine like terms to solve the equation.

If the like terms are on the same side of the equation, add or subtract them as indicated.

If the like terms appear on different sides of the equation, combine them using inverse operations.

$$5x + 2x - 6 = 12$$
$$7x - 6 = 12$$
$$7x = 18$$
$$x = \frac{18}{7}$$
$$x = 2\frac{4}{7}$$

$$\begin{array}{r} a + 12 = 14 - a \\ + a \qquad\qquad + a \\ \hline 2a + 12 = 14 \\ 2a = 2 \\ a = 1 \end{array}$$

$$cb = \frac{86 - 5}{cb}$$
$$(cb)(cb) = \frac{cb(86 - 5)}{cb}$$
$$c^2b^2 = 86 - 5$$
$$c^2b^2 = 81$$
$$cb = 9 \text{ or } -9$$

To solve an algebraic problem, always start by working within parentheses. The "order of operations" is to first evaluate powers and roots, then perform multiplication or division, and then perform addition or subtraction.

Tip: This is called the "order of operations":
1. Work within parentheses.
2. Evaluate powers or roots.
3. Multiply or divide.
4. Add or subtract.

PRACTICE

Solve and check each equation below.

1 $6y - 2y = 84$

2 $5a + 2a + 3 = 24$

3 $3x = 2x + 12$

4 $6f = 2.5 + f$

5 $9r + 17 = 6r + 32$

6 $6n - 9 = 2n + 7$

7 $\frac{t + 2t^2}{t} = 43$

8 $2n - 13 = 3n$

9 $3x + 6x - 11 = 35 - 10$

10 $a - \left(\frac{1}{2}\right)a = 3$

11 $\left(\frac{1}{2}\right)b + \left(\frac{1}{4}\right)b = 20 + 4$

12 $6n - 12 = 2n - 4$

Solving Inequalities

An **inequality** is a statement that two expressions are not equal. It consists of two expressions joined by one of the signs <, >, ≤, or ≥.

Sign	Inequality	Solution	Interpretation of Solution
less than (<)	$-3 + n < 0$	$\begin{aligned} -3 + n &< 0 \\ +3 \qquad &+3 \\ n &< 3 \end{aligned}$	n is any number less than 3.
greater than (>)	$x^2 > \dfrac{x}{8}$	$x^2 > \dfrac{x}{8}$ $x > \dfrac{1}{8}$	x is any number greater than $\dfrac{1}{8}$.
less than or equal to (≤)	$-m \le 7 + 3$	$\begin{aligned} -m &\le 7 + 3 \\ -m &\le 10 \\ m &\ge -10 \end{aligned}$	m is any number greater than or equal to negative 10.
greater than or equal to (≥)	$\dfrac{a}{8} \ge \dfrac{4}{5}$	$\dfrac{a}{8} \ge \dfrac{4}{5}$ $(8)\left(\dfrac{a}{8}\right) \ge (8)\left(\dfrac{4}{5}\right)$ $a \ge \dfrac{32}{5}$ $a \ge 6\dfrac{2}{5}$	a is any number greater than or equal to $6\dfrac{2}{5}$.

You solve an inequality just as you solve an equation. You use inverse operations to isolate the unknown on one side of the equation, then simplify the other side. There is one exception, as shown above in the step from $-m \le 10$ to $m \ge -10$: If you multiply or divide by a negative number, you *reverse* the order of the inequality.

PRACTICE

Solve each inequality below. Reduce all fractions to simplest terms.

1 $a - 5 > 22$

2 $b + 15 < -10$

3 $4x > 15$

4 $3u + 5 \ge 17$

5 $8y < 18 - y$

6 $\left(\dfrac{1}{2}\right)n + 17 \le 25$

7 $\dfrac{a-5}{10} < 20$

8 $\left(\dfrac{2}{3}\right)y < 36$

9 $x^2 - 5 < 11$

10 $5a - 3a + 3 > 17$

11 $5y + y \le 20$

12 $4b(b) < 16$

Using Formulas

The equation $i = prt$ shows how simple interest (i) is related to principal (p), rate (r), and time (t). The equation is an example of a **formula,** which is a rule given in algebraic form.

> To use a formula, you substitute known values for all but one of the letters in the formula. Then you solve the equation for the unknown value.

Example:

Your brother borrows $2,000 from you, then pays it back 2 years later with $200 interest. What was the rate of interest you charged your brother for the loan?

$2,000 is the principal p
$200 is the interest i
2 years is the time t

Solution:
Substitute those values into the formula. Then solve for r.

$$i = prt$$
$$200 = (2,000)(r)(2)$$
$$200 = 4,000r$$
$$\frac{200}{4000} = r$$
$$r = 0.05$$
$$r = 5\%$$

PRACTICE

Solve each problem using the given formula.

1   Josh puts $1,500 into an account that earns 6% interest each year. How many years will it take him to earn $450 in interest? (Use $i = prt$.)

2   A turtle crawls at an average rate r of 2 feet a minute. At that rate, what distance d will it travel in a time t of 10 minutes? (Use $d = rt$.)

3   If Sonia hikes at an average speed of 2 miles an hour, how many hours will it take her to hike 7.4 miles? (Use $d = rt$.)

4   A grocer buys 104 flats of strawberries at a rate of $14.00 per flat. How much does he pay? (Use $c = rn$, where c is the total cost, r is the unit rate, and n is the number of units.)

Solving Word Problems Using Algebra

Solve each problem below.

1 Wanek's Department Store charges $25.00 to deliver a piece of furniture, plus $0.50 for every mile over 25. Which of these equations could the store use to figure the charge (c) for each delivery? (Let m stand for the number of miles between the store and the customer's house.)

    A $\$25.00 + \$0.50m = c$
    B $m(\$25.00 + \$0.50) = c$
    C $\$0.50(\$25.00 + m) = c$
    D $\$25.00 + \$0.50(m - 25) = c$

2 There are 450 women and 2,250 men attending Santa Clara College. The college wants to have a 3:1 ratio of men to women. To get that ratio, how many more women must they recruit?

3 Rita works $7\frac{1}{2}$ hours a day. It takes her $\frac{1}{3}$ hour to complete a haircut and 1 hour to give a permanent. Tomorrow, Rita has three appointments for permanents. How many haircuts can she complete? (Do not count fractions of a haircut.)

4 Last year, $42,000,000 in donations were made to United Charities. One-half of that money was used to pay salaries and other administrative costs, and one-third of the remaining money was used to pay for mailings. How much money was left?

5 Delicia and Dawn run a cleaning business. Delicia does $\frac{2}{3}$ of the work and keeps $\frac{2}{3}$ of the profits. Last year she made $32,000. What were the total profits of the business?

6 Four-fifths of the sum of three and a number n is 64. What is n?

7 Scott makes $16.00 an hour delivering packages. Last night he worked for $5\frac{1}{4}$ hours and spent $3.75 of his own money on gas. How much money did he make?

8 The difference between a positive number and 7 is divided by 12. The quotient is 4. What is the number?

Algebra Skills Practice

Circle the letter for the correct answer to each problem.

1

$7x + x =$ _____

- **A** $7x^2$
- **B** $7x$
- **C** $8x$
- **D** 7
- **E** None of these

2 $3(-x + 2y) - 15 =$ _____

- **F** $3x + 6y - 15$
- **G** $-3x + 5y - 15$
- **H** $-3x + 6y - 15$
- **J** $-3x + 2y - 15$
- **K** None of these

3

$\dfrac{3ab^2 - 2ab}{ab} =$ _____

- **A** $3b - 2$
- **B** 1
- **C** $\dfrac{3}{b}$
- **D** $3ab^2 - 2$
- **E** None of these

4 $5y^2 - 3y^2 + 2y^2 =$ _____

- **F** 0
- **G** $5y^2$
- **H** y
- **J** $5y^5 - 5y^0$
- **K** None of these

5 $-x(3x - 3m) =$ _____

- **A** $3x^2 + 3mx$
- **B** $-3x^2 + 3mx$
- **C** $-3x^2 - 3mx$
- **D** mx^3
- **E** None of these

6 Every day before work Rodney buys a double latte for $2.75, but every tenth latte he gets is free. Which of these number sentences could Rodney use to figure out how much he spends on lattes in a month? (Assume there are 21 workdays in a month.)

- **F** $\$2.75 \times 21 = \square$
- **G** $\$2.75(21) - \dfrac{1}{10} = \square$
- **H** $\dfrac{1}{10} \times \$2.75 \times 21 = \square$
- **J** $\dfrac{9}{10} \times 21 \times \$2.75 = \square$

7 Lori plants a 3-foot-tall maple tree that grows 5 feet a year. At the same time, she plants a 6-foot-tall oak that grows 2 feet a year. Which of these equations could you use to figure out how many years (y) it will take the maple tree to catch up to the oak in height?

- **A** $5x = 2y$
- **B** $\dfrac{6-3}{y} = \dfrac{2}{5}$
- **C** $3 + 5y = 6 + 2y$
- **D** $y(3 + 5) = y(6 + 2)$

8 Which of these equations represents the relationship that 10 times the sum of a number and 12 is 240?

- **F** $10n + 12 = 240$
- **G** $10(n + 12) = 240$
- **H** $n + 12 = 10(240)$
- **J** $10(n - 12) = 240$

9 Tom buys 36 square feet of brick. If he uses the brick to make a square patio, how long will each side be?

- **A** 18 feet
- **B** 9 feet
- **C** 6 feet
- **D** 12 feet

10 $3(a + b) + (a - b) =$ _____

    F $4a$
    G $4a\ 4b$
    H $4a + 2b$
    J $4a + 4b$
    K   None of these

11 $\dfrac{10x^2 + 2x}{2x} =$ _____

    A $5x$
    B $5x + 1$
    C $10x^2$
    D $4x$
    E   None of these

12 $(9ab^2c^2)(bc^3) =$ _____

    F $9ab^2c^6$
    G $9ab^3c^6$
    H $9ab^3c^5$
    J $(9abc)^3$
    K   None of these

13 $6xy + 3y(x^2 - 2x) =$ _____

    A 0
    B $12xy$
    C $3x^2y$
    D $3xy$
    E   None of these

14 $p\left(3n + \dfrac{1}{p}\right) =$ _____

    F $3np + 1$
    G $3n + 1$
    H $3np + 1/p$
    J $3n$
    K   None of these

15 $\dfrac{5(2x + x)}{x} =$ _____

    A $15x^2$
    B 15
    C 10
    D $10x$
    E   None of these

16 Which answer choice shows how to find the next number in this pattern?

    1, 1, 2, 6, 24, _____

    F $24 + 18 =$ _____
    G $24 \times 4 =$ _____
    H $24 + (6 \times 4) =$ _____
    J $24 \times 5 =$ _____

17 There are 100 jelly beans in a bag. Martha needs 25 jelly beans for each of the 23 children in her class. Which of these number sentences could she use to figure out how many bags of jelly beans she needs?

    A $(100 \times 25) \div 23 =$ _____
    B $100 \times 25 \times 23 =$ _____
    C $100 \div (25 \times 23) =$ _____
    D $(25 \times 23) \div 100 =$ _____

18 Which answer choice shows how to find the next number in the pattern?

    0.0009018, 0.09018, 9.018, 901.8, ____

    F $901.8^2 =$ _____
    G $901.8 \times 100 =$ _____
    H $901.8 \times 10 =$ _____
    J $901.8 \div 100 =$ _____

19 Which of these equations represents the relationship that one-third of a number n minus six equals forty-two?

    A $\dfrac{n - 6}{3} = 42$

    B $n - 6 = \dfrac{42}{3}$

    C $\dfrac{n}{3} - 6 = 42$

    D $n - \dfrac{6}{3} = 42$

Measurement

Reading a Scale

A number line on a measurement tool is called a **scale.**
You can estimate how close the pointer is to the lower or higher label.

You can estimate the reading on a scale.

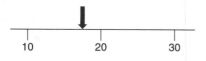

This pointer is about $\frac{4}{5}$ of the way from 10 to 20. The reading on the scale is about 18.

If the scale has tick marks, count the number of spaces formed between numbers. Then figure out what each tick mark is worth.

In the scale above there are 5 tick marks between 50 and 100, so each tick mark represents 10 units.

PRACTICE

Find each measurement below.

1 The pointer on a scale is $\frac{3}{5}$ of the way from 15 to 20.

_____

2 The pointer is about $\frac{2}{3}$ of the way from 0 to 100.

_____

3 The pointer is about $\frac{1}{5}$ of the way from 60 to 80.

_____

4

- 2 cup

- 1 cup

_____

5

50   75

_____

6

0
35   5
30   10
25   15
20

_____

7

50°

0°

What measurement is shown on the scale?

_____

8 On a scale, the interval between 50 and 100 is divided into 5 spaces. What is each tick mark worth?

_____

9 On a scale, the interval between 25 and 35 is divided into 5 spaces. What is each tick mark worth?

_____

10

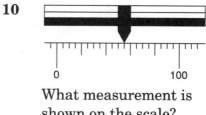

0   100

What measurement is shown on the scale?

_____

Standard Units of Measure

One system of measurement uses **standard units.**

Standard Units of Measurement

Temperature degrees Fahrenheit (°F)	Normal body temperature is 98.6°F. Water boils at 212°F. Water freezes at 32°F.

Length
    12 inches (in.) = 1 foot (ft)
        3 feet = 1 yard (yd)
    5,280 feet = 1 mile (mi)
	An inch is about the length of a straight pin.
	A foot is about the length of a man's foot.
	A yard is about the length of your arm.
	A mile is 8–10 city blocks.

Weight
    1 pound (lb) = 16 ounces (oz)
    1 ton (T) = 2,000 lb
	A pencil weighs about 1 ounce.
	An eggplant weighs about 1 pound.
	A car weighs about 1 ton.

Capacity
        1 pint = 2 cups
    1 quart = 4 cups (or 2 pints)
    1 gallon = 4 quarts (or 16 cups)
	An ice cream dish holds 1 cup.
	A mug holds about 1 pint.
	A thin milk carton holds 1 quart.
	A plastic milk jug holds 1 gallon.

Time
    1 minute (min) = 60 seconds (sec)
        1 hour (hr) = 60 min
            1 day = 24 hr

PRACTICE

Circle the letter for the best estimate for each measurement.

1   the temperature of frozen ice cream

   A   70°F
   B   30°F
   C   48°F

2   the length of a shoelace

   F   3 inches
   G   3 feet
   H   14 inches

3   the weight of an audio tape

   F $\frac{1}{2}$ pound

   G $\frac{1}{3}$ ounce

   H   3 ounces

4   the capacity of a top hat

   A $1\frac{1}{2}$ cups

   B $\frac{1}{2}$ gallon

   C $\frac{1}{2}$ quart

Circle the *larger* measurement in each pair.

5   8 inches   1 foot

6   5 feet   1 yard

7   10 ounces   1 pound

8   10 ounces $\frac{1}{2}$ pound

9   $\frac{1}{2}$ cup   1 pint

10   10 cups   1 gallon

11   1 gallon   2 quarts

Using a Standard Ruler

Most rulers marked in inches use several different types of tick marks:

- The longest tick marks are numbered. They divide the ruler into inches.
- The next longest tick marks divide the inches into halves.
- The fourths of an inch are represented by slightly shorter tick marks.
- The eighths of an inch are represented by even shorter tick marks.
- The shortest tick marks may represent sixteenths or thirty-seconds of an inch.

PRACTICE

On a standard ruler, find the tick marks that divide each inch into fourths. Use those tick marks to measure each line below to the nearest $\frac{1}{4}$-inch. Reduce all fractions to lowest terms.

1 ──────────────────

_____ inches

2 ──────────────

_____ inches

3 ───

_____ inches

4 ─────────────

_____ inches

5 ──

_____ inches

Find the marks on your ruler that divide each inch into eighths. Use those tick marks to measure each line below to the nearest $\frac{1}{8}$-inch. Reduce all fractions to simplest terms.

6 ─────────

_____ inches

7 ──────────────────

_____ inches

8 ────────────────────────

_____ inches

9 ───

_____ inches

10 ────

_____ inches

Measure each line below to the nearest $\frac{1}{16}$-inch. Reduce all fractions to simplest terms.

11 ——

_____ inches

12 ——

_____ inches

13 ——

_____ inches

14 ———————

_____ inches

15 ——

_____ inches

16 ———————

_____ inches

Find _two_ of the objects listed below. Use the lines provided to record their height, width, and length to the nearest $\frac{1}{4}$-inch.

17 a phone book (height and width only)
18 a 6.75-ounce juice box
19 a 200-square-foot box of aluminum foil
20 a 20-bag box of gallon-sized storage bags
21 a 7.25-ounce box of macaroni and cheese, with powdered cheese
22 a 300-bag box of sandwich bags
23 a 100-square-foot box of plastic wrap
24 a VHS video tape

Object 1: _____

_____ in. by _____ in. by _____ in.

Object 2: _____

_____ in. by _____ in. by _____ in.

Measure the height, width, and length of _two_ objects listed below. Give your answers to the nearest $\frac{1}{8}$-inch.

25 a 16-ounce box of spaghetti
26 a $9\frac{1}{2}$- or 10-ounce box of crackers
27 a 1.7-ounce box of teabags
28 a 9-volt battery, including the terminals
29 a 20-bag box of quart-size storage bags
30 a 2-pound box of baking soda

Object 1: _____

_____ in. by _____ in. by _____ in.

Object 2: _____

_____ in. by _____ in. by _____ in.

Using a Standard Ruler

Converting Within the Standard System

One way to convert between units is to use a proportion. Put like units in the numerator and like units in the denominator.

Problem: How many feet are in 50 inches?

Solution:

Set up a proportion. $\frac{12 \text{ in.}}{1 \text{ ft}} = \frac{50 \text{ in.}}{? \text{ ft}}$

Cross multiply. $? \times 12 = 1 \times 50$

Divide both sides by the number 12.

$\frac{? \times 12}{12} = \frac{1 \times 50}{12}$

$? = \frac{50}{12} = 4\frac{1}{6}$

Another way to convert between units is to use multiplication or division.

To change a measurement to smaller units, you can multiply.	To change a measurement to larger units, you can divide.
Examples:	**Examples:**
To change feet to inches, multiply by 12.	To change feet to yards, divide by 3.
To change pounds to ounces, multiply by 16.	To change pounds to tons, divide by 2,000.

PRACTICE

Convert each measurement below to the given unit.

1. 9 feet = _____ yards

2. 2.5 feet = _____ in.

3. 54 in. = _____ ft

4. $\frac{1}{4}$ ft = _____ in.

5. $\frac{1}{4}$ lb = _____ oz

6. $\frac{3}{4}$ foot = _____ inches

7. $\frac{1}{2}$ yard = _____ feet

8. 1 lb, 3 oz = _____ oz

9. 1 hr, 10 min = _____ min

10. 3 yd, 2 ft = _____ feet

11. 12 ft, 3 in. = _____ in.

12. 1 gal, 2 qt = _____ qt

13. 15 in. = _____ ft, _____ in.

14. 35 oz = _____ lb, _____ oz

15. 3 gallons = _____ quarts

16. $\frac{1}{2}$ gal = _____ qt

17. How many cups fill an 8-gallon water cooler?

    _____

18. Rina walks 4.3 miles. How many feet is that?

    _____

19. The electric company promises to restore power within 48 hours. How many days is that?

    _____

20. One box of cereal weighs 14 oz. What fraction of a pound is that?

    _____

The Metric System

The **metric system of measurement** uses **metric units.** Most countries around the world use the metric system.

Metric Units of Measurement

Temperature Degrees Celsius (°C)	Water boils at 100°C. Water freezes at 0°C. Normal body temperature is 37°C.

Length
    1 meter (m) = 1,000 millimeters (mm)
                = 100 centimeters (cm)
    1 cm = 10 mm
1 kilometer (km) = 1,000 m
	A needle is about 1 mm wide.
	A kindergarten student is about 1 meter tall.
	Your little finger is about 1 centimeter wide.

Weight 1 gram (g) = 1,000 milligrams (mg) 1 kilogram (kg) = 1,000 g	A needle weighs about 1 milligram. A peanut weighs about 1 gram. A city telephone book weighs about 1 kg.
Capacity 1 liter (L) = 1,000 milliliters (mL) 1 kiloliter (kL) = 1,000 L	A large plastic soda bottle holds 2 liters. A dose of cough medicine is about 1 mL. A septic tank holds about 2 kiloliters.

The metric system is based on **powers of ten.** (The powers of ten are 1, 10, 100, 1,000, 10,000, and so on.) The basic units are **meter, gram,** and **liter.** Any measurement that begins with *milli-* is one-thousandth of the basic unit. Any measurement that begins with *kilo-* is one thousand times the basic unit.

PRACTICE

Fill in each blank.

1   A cherry weighs about 1 _____.

2   An ashtray holds about 10 _____.

3   A magazine weighs about $\frac{1}{5}$ _____ .

4   The width of your ear is about 5 _____.

5   A potato weighs about 20 _____.

6   $\frac{1}{2}$ gram = _____ mg

7   2 cm = _____ mm

8   4,200 mg = _____ g

9   $\frac{1}{4}$ meter = _____ cm

10   500 mm = _____ m

11   10,000 mm = _____ cm

12   $\frac{3}{4}$ kL = _____ L

13   15,000 mL = _____ L

14   2L, 150 mL = _____ mL

15   12g, 700 mg = _____ mg

16   $2\frac{1}{2}$ meters = _____ cm

17   5.3 kL = _____ liters

       The Metric System

Using a Metric Ruler

Metric rulers are used in most parts of the world. Most metric rulers use two units. The larger units are centimeters (cm). The smaller units are millimeters (mm). There are 10 millimeters in every centimeter.

Use a metric ruler to measure each line below.

1 ————————————————

_____ centimeters

2 ————————

_____ centimeters

3 ——

_____ centimeters

4 ——————————

_____ centimeters, _____ millimeters

5 ——————————

_____ centimeters, _____ millimeters

6 ——————————

_____ centimeters, _____ millimeters

The objects listed below are difficult to measure because they have rounded or irregular shapes. Measure the *widest and tallest* parts of four objects listed. Give your answers to the nearest half centimeter.

7   a large coffee can (966 g or 34.5 oz)
8   a plastic canister for 35-mm film
9   a 13-oz coffee can
10   a can of vegetables (14–16 oz or 392–448 g)
11   a 355 mL (12 oz) bottle of saline solution
12   a 210 g (7.5 oz) can of unbaked biscuits
13   a 1-dozen carton of eggs
14   a 230 g (8.2 oz) tube of toothpaste
15   a 450 mL (16 oz) bottle of rubbing alcohol
16   a 7 g ($\frac{1}{4}$ oz) package of dry yeast

Object 1: _____

Measurements: _____ cm by _____ cm

Object 2: _____

Measurements: _____ cm by _____ cm

Object 3: _____

Measurements: _____ cm by _____ cm

Object 4: _____

Measurements: _____ cm by _____ cm

Comparing Standard and Metric Units

You should have a general idea of how standard and metric units compare. The symbol "≈" means "is approximately equal to."

1 inch ≈ 2.5 cm 1 meter ≈ 1.1 yards 1 kilometer ≈ 0.6 mile	1 kilogram ≈ 2.2 lb 1 ounce ≈ 28 g	1 liter ≈ 1.1 qt 1 kiloliter ≈ 275 gal

On the previous pages you converted measurements *within* the standard system and *within* the metric system. You can use the same process to get equivalent measurements *between* the two systems.

Problem: 40 cm ≈ __?__ inches

Solution:

Set up a proportion. $\dfrac{1 \text{ in.}}{2.5 \text{ cm}} = \dfrac{? \text{ in.}}{40 \text{ cm}}$

Cross multiply. $2.5 \times ? = 1 \times 40$

Divide by 2.5. $\dfrac{2.5 \times ?}{2.5} = \dfrac{40}{2.5}$

$? = 16$

PRACTICE

Circle the larger measurement in each pair.

1 1 meter 1 inch

2 1 yard 1 meter

3 1 gallon 1 liter

4 1 kilometer 1 mile

5 1 ounce 1 gram

6 1 inch 1 centimeter

7 1 kiloliter 1 gallon

8 1 liter 1 quart

9 1 kilogram 1 pound

Fill in each blank. Round all answers to the nearest tenth.

10 18 yards ≈ _____ meters

11 23 grams ≈ _____ ounces

12 12 mi ≈ _____ km

13 10 quarts ≈ _____ liters

14 2 yards ≈ _____ meters

15 8 lb ≈ _____ kg

16 25 cm ≈ _____ in.

Adding and Subtracting Mixed Measurements

When two measurements are in the same units, you can add them or subtract them.

To add mixed units:

1. Add the smaller units.
2. Add the larger units.
3. Simplify your answer by converting as many of the smaller units as possible into the larger units.

```
  1 hour    15 minutes
+ 2 hours   50 minutes
  3 hours   65 minutes
```

= 3 hours + 60 minutes + 5 minutes
= 3 hours + 1 hour + 5 minutes
= 4 hours, 5 minutes

To subtract mixed units:

1. Subtract the small and large units separately.
2. Simplify your answer.

In a subtraction problem, you may have to borrow.

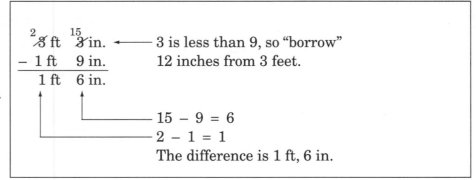

3 is less than 9, so "borrow" 12 inches from 3 feet.

15 − 9 = 6
2 − 1 = 1
The difference is 1 ft, 6 in.

PRACTICE

Simplify each measurement.

1   1 hr, 90 min

2   2 ft, 15 in.

3   11 yd, 9 ft

4   5 gal, 6 qt

Add or subtract the measurements below. Simplify your answers.

5   3 hr, 50 min
   + 2 hr, 20 min

6   10 lb, 12 oz
   + 5 lb, 9 oz

7   13 yd, 2 ft
   + 16 yd, 1 ft

8   18 ft, 12 in.
   + 5 ft, 17 in.

9   3 hr, 45 min
   +   30 min

10   3 qt, 2 cups
   + 1 qt, 3 cups

11   5 lb, 9 oz
   − 3 lb, 3 oz

12   10 yd, 1 ft
   − 2 yd, 3 ft

13   6 ft, 3 in.
   − 4 ft, 5 in.

14   5 gal, 1 qt
   −   3 qt

15   7 hr, 10 min
   − 4 hr, 25 min

16   2 pt
   − 1 pt, 1 cup

Multiplying and Dividing Mixed Units

When you multiply mixed units by a whole number, multiply the large and small units separately.

Then simplify your answer.

To divide mixed units:
- ◆ Convert them into the smaller unit.
- ◆ Multiply or divide as indicated.
- ◆ Simplify your answer if necessary.

Example:

1 ft 3 in.	
× 12	

12 ft 36 in. = 12 ft + 3 ft = 15 ft

Problem:

How many full pieces will you get if you divide a ribbon 12 yards long into sections of length 1 yard, 2 feet?

Solution:

Divide 12 yards by 1 yard, 2 feet after converting all the measurements to feet.

12 yd = 36 ft

1 yd, 2 ft = 3 ft + 2 ft = 5 ft

Divide 36 by 5.

$$5\overline{)36} \quad \frac{7}{}$$

35

1

The quotient is 7 r 1. You will get 7 full pieces of ribbon.

PRACTICE

Solve each problem below, then simplify your answer. Round all answers to the nearest tenth.

1   14 ft 6 in.
      ×   12

2   10 lb 8 oz
      ×   5

3   5 yd 2 ft
      ×   6

4   What is 2 feet, 6 inches divided by 10?

5   What is 3 lb, 2 oz ÷ 5 ?

6   What is 7 yards, 3 feet divided by 4 feet?

7   Dimitri has 5 gallons, 8 cups of egg nog. How many $\frac{2}{3}$-cup glasses will this fill?

8   Antonio has 6 rolls of tape, and each roll contains 12 feet, 6 inches of tape. How much tape does he have?

9   How many 12-minute sessions will fit into 1 hour, 12 minutes?

10   Franz has 1 kilogram, 250 grams of flour. His pancake recipe calls for 250 grams of flour. How many batches of pancakes can he make?

Finding Perimeter

The total length of the sides of a figure is called its **perimeter.**

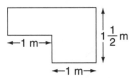

The length of the top and bottom is 2 meters, and the length of the left and right edges is $1\frac{1}{2}$ meters. The perimeter is

$2 + 1\frac{1}{2} + 2 + 1\frac{1}{2} = 7$, or 7 meters.

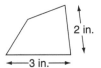

We do not know the lengths of the two unmarked sides. So we do not have enough information to calculate the perimeter of the figure.

PRACTICE

Find the perimeter of each shape. *Hint:* **The symbol " is an abbreviation for inches. The symbol ' is an abbreviation for feet.**

1 A rectangular room is 21 feet long by 11 ft, 2 in. wide.

Perimeter: _____

2

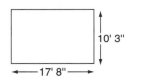

Perimeter: _____

3

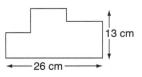

Perimeter: _____

4

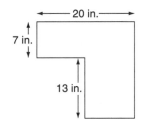

Perimeter: _____

5

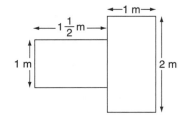

Perimeter: _____

6 This figure is made up of three squares, each with sides 3 mm long.

Perimeter: _____

7

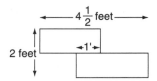

Perimeter: _____

8

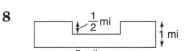

Perimeter: _____

9 The perimeter of a square is 80 m. How long is each side?

_____

10 The perimeter of a rectangle is 9 feet. The length is $2\frac{1}{2}$ feet long. How long is the width? (Try drawing this figure.)

_____

Finding the Circumference of a Circle

The distance around a circle is called its **circumference.**

You can find the circumference of a circle if you know the distance across the circle, or its **diameter.**

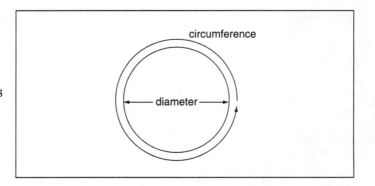

A formula for the circumference of a circle is $C = \pi d$, where C is the circumference, d is the diameter of the circle, and π is a number approximately 3.14 or $3\frac{1}{7}$.

PRACTICE

In problems 1–5, find each circumference. Round all answers to the hundredths place. If you do not have enough information to find the circumference, write "Cannot tell."

1

—4 in.—

Circumference: _____

2

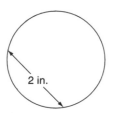

2 in.

Circumference: _____

3

—20 cm—

Circumference: _____

4

—6 mm—

Circumference: _____

5

•—5"—

Circumference: _____

6 The curve below is one quarter of a circle.

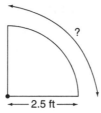

?

—2.5 ft—

What is the length of the curve? _____

7 Elaine's circular hot tub is 8 feet in diameter. How much rubber tubing would she need to seal the joint around the bottom of the tub?

_____

8 Carl has a circular flower bed $2\frac{1}{2}$ feet across through the center. How many feet of edging does he need to go around the bed?

_____

116

Finding Area

The amount of a flat region inside a figure is called its **area.**

Figure	Area Formula	Definition of Terms	Example
square or rectangle	$A = lw$	l = length w = width	3 m / 2 m Area = 3 m × 2 m
triangle	$A = \frac{1}{2}bh$	b = base h = height	6 ft / 10 ft Area = $\frac{1}{2}$ × 10 ft × 6 ft
circle	$A = \pi r^2$	π is about 3.14 or $\frac{22}{7}$ r = radius, which is the distance from the circle to its center; the radius is half the diameter.	2 in. Area = 3.14 × 2 in. × 2 in.

Area is measured in *square units* (units^2). A square yard is a square whose sides are each 1 yard long. A square foot is a square with 1-foot sides.

PRACTICE

Use this diagram for problems 1–3.

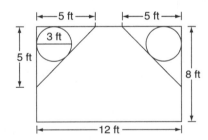

1   the area of each circular planter (Round your answer to the nearest whole number.) _____

2   the total area of the deck _____

3   the area of the white space in the middle of the deck _____

This diagram is part of the plans for a window box. Use the diagram for questions 4–6.

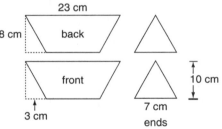

4   the area of the front _____

5   the area of each end _____

6   To paint the outside of this box, including the back, you would need enough paint to cover what area? _____

Finding Volume

The amount of space inside a three-dimensional figure is called its **volume.**

Figure	Volume Formula	Definition of Terms	Example
rectangular solid (box, cube)	$V = lwh$	l = length w = width h = height	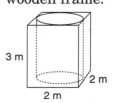 Volume = 60 × 40 × 40
cylinder	$V = \pi r^2 h$	r = radius h = height $\pi \approx 3.14$ or $\frac{22}{7}$	Volume = 3.14 × 3 × 3 × 6

Volume is measured in *cubic units,* or units³. One cubic foot is a cube one foot long on each side. One cubic centimeter is a cube 1 centimeter long on each side.

PRACTICE

Find the volume of each figure below.

1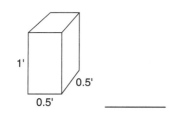
0.5 in.   10 in.   5 in.

_____

2
1'   0.5'   0.5'

_____

3 A glass with a radius of 2 inches and a height of 9 inches is half full of wine. How much wine is there?

_____

4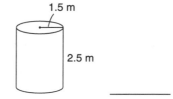
1.5 m   2.5 m

_____

5 A sugar cube is 2 cm on every side. What is its volume?

_____

6 A pail is 12 inches wide (so r = 6 in.) and 10 inches tall. How many times would you have to fill the pail to move 5,600 cubic inches of dirt?

_____

7 This is a cylindrical garbage can inside a wooden frame.

3 m   2 m   2 m

What is the volume of the frame minus the volume of the garbage can? (Round to the nearest tenth.)

_____

Changes in Time

For the problem below, imagine how the hands on the clock move. Count the number of *complete* hours that pass. Then count the minutes that pass.

> **Problem:**
> Eric started work at 8:25. He finished at 4:10. How long did he work?
>
> **Count the hours:** The number of *complete* hours from 8:25 to 3:25 is 7.
> **Count the minutes:**
> From 3:25 to 4:10 is 45 minutes.
>
> **Answer:** 7 hours, 45 minutes

For the problem below, you want to find when you should start a task in order to finish at a particular time. To do this, count the change in hours first. Then add the change in minutes.

> **Problem:**
> It takes 1 hour, 15 minutes to drive to a doctor's office. You have a 10:00 appointment. When should you leave?
>
> *Hint:* You need a time *before* 10:00, so count back from 10:00.
>
> **Count the hours:**
> 1 hour before 10:00 is 9:00.
> **Count the minutes:**
> 15 minutes before 9:00 is 8:45.
> **Answer:** You should leave at 8:45.

Pay close attention to any mention of A.M. (morning) and P.M. (afternoon and evening). From 9:00 A.M. to 10:00 A.M. is 1 hour, but from 9:00 A.M. to 10:00 P.M. is 13 hours.

PRACTICE

Fill in the blank for each problem.

1   What time is 6 hours and 10 minutes before 12:05 noon? (Include A.M. or P.M.) _____

2   The drive to work takes 43 minutes. What time must you leave in order to get to work by 9:00 A.M.? _____

3   A concert starts at 7:22 P.M. and ends at 9:15 P.M. How long did it last? _____

4   It is 3 hours earlier in California than it is in New York. A New Yorker wants to call a friend at 6:30 P.M. California time. At what "New York time" should she call? _____

5   You plan to start work at 8:30 A.M. You will spend 1 hour writing a report, $1\frac{1}{2}$ hours in a meeting, and 3 hours checking data. When will you be finished? _____

6   It takes 14 minutes to cook one batch of cinnamon rolls. You put your first batch in the oven at 7:10. When will your third batch be finished? _____

7   Anne began reading a 300-page novel at 8:15 one evening. She finished it at 2:15 the next morning. How many pages did she read per hour? _____

Measurement Skills Practice

Circle the letter for the correct answer to each problem.

1 Which of these is the same amount as $5\frac{1}{2}$ gallons of apple cider?

(1 gallon ≈ 3.785 liters)

A 20.8 liters
B 1.4 liters
C 2.08 liters
D 14 liters

2 During a typical day, Frederick spends $7\frac{1}{2}$ hours sleeping, $7\frac{1}{2}$ hours working, and 1 hour shopping or cleaning. What fraction of the day is left for other activities?

F $\frac{1}{4}$ **H** $\frac{1}{6}$

G $\frac{1}{5}$ **J** $\frac{1}{3}$

3 A movie is 1 hour, 56 minutes long. Claire begins watching the movie at 5:14 P.M. When will she finish watching?

A 7:00 P.M.
B 6:10 P.M.
C 7:10 P.M.
D 6:70 P.M.

4 How many yards is 114 feet?

F 38
G 9.5
H 36.8
J 11.4

5 What is the area of this figure?

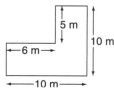

A 100 m²
B 130 m²
C 70 m²
D 20 m²

6 This vase is 30 centimeters tall and has a radius of 3 centimeters.

Latrice wants to buy enough marbles to fill the vase $\frac{1}{3}$ full. About how many cubic centimeters of marbles does she need?

F 60 cm³
G 90 cm³
H 180 cm³
J 270 cm³

7 A box 10 cm wide, 12 cm tall, and 20 cm long is filled with a product that weighs 13 grams per cubic centimeter. What is the weight of the product inside the box?

A 31 kilograms, 200 grams
B 2 kilograms, 400 grams
C 184 kilograms
D 312 grams

8 Mark and Buddy spend 6 hours resurfacing the cabinets in a client's kitchen. They spent $510 on materials, and the client paid $750 for the job. If Dominic and Buddy split the profits evenly, how much will each worker have made per hour?

F $20.00
G $40.00
H $62.50
J $120.00

Use the following information to do Numbers 9 through 11.

Jordan is recovering his kitchen floor. The floor is 15 feet wide and 22 feet long. The tiles are 6 inches by 6 inches.

9   In which of these number sentences is x the number of tiles Jordan needs?

A $\dfrac{(15 \times 12)(22 \times 12)}{6 \times 6} = x$

B $\dfrac{15 \times 22}{6 \times 6} = x$

C $\dfrac{15 \times 22}{6} = x$

D $15 \times 22 = x$

10   One can of tile adhesive covers 100 square feet. How many cans does Jordan need?

F   1
G   2
H   3
J   4

11   After 30 minutes, Jordan has covered $\frac{1}{5}$ of the floor. At this rate, how much longer will it take to finish the job?

A   1 hour
B   2 hours
C $2\frac{1}{2}$ hours
D   3 hours

12   Between 9:00 P.M. and 1:00 A.M. one night, the temperature in Bismarck dropped from 10°F to –2°F. During that period, what was the average drop in temperature per hour?

F   2.5°F
G   3°F
H   2°F
J   3.5°F

13   Viola bought 400 quilting squares. Each one is a 4-inch by 4-inch square. Viola plans to sew all the squares into a square quilt. How large will the quilt be?

A   1,600 in.²
B   6,400 in.²
C   10,400 in.²
D   160,000 in.²

14   About how many inches are there in 10 centimeters?

F   2   H   5
G   10   J   15

15   How many square inches are in one square foot?

A   12   C   144
B   24   D   9

16   At the Little Tots Preschool, $\frac{1}{4}$ of every hour is spent in free play. If Jenny's son spends 6 hours at the school, how much time will he spend in free play?

F   1 hour, 15 minutes
G $1\frac{1}{2}$ hours
H   1 hour, 40 minutes
J   2 hours

Geometry

Visualizing

A key skill in geometry is the ability to visualize a figure. You should be able to imagine what a solid object looks like if you view it from different angles. To develop that skill, draw pictures of the figures in problems whenever you can.

PRACTICE

As you read each problem on this page, draw or imagine the figure it discusses. Then use that figure to solve the problem.

1

The diagram above shows the parts of a window box. Which choice below shows what the window box will look like once it is assembled?

2

The drawing above shows a solid cylinder, which is a shape like a tin can. If you cut this cylinder in half along the dotted line, what shape would the cut end have?

F   **G**   **H**   **J**

3

This figure is a cube, which is a box whose sides are all the same length. How many sides does a cube have?

_____

4

This figure is a rectangular solid. It is a box whose length, width, and height may have different measures. How many small cubes have been put together to create this rectangular solid?

_____

5 Circle the letter for the figure that would hold more water.

Cylinder   Cone

A   **B**

Basic Concepts

Geometry is the study of shapes. The ideas below are the building blocks used to create and describe shapes.

Basic Ideas in Geometry

Name	Meaning	Diagram	Symbol
point	a dot; a single location	•A	a single letter, such as A or B
line	a straight, connected set of points; it extends in two directions	A B	two letters with a double arrow above, such as $\overleftrightarrow{AB}$
line segment	the part of a line that is between two endpoints	A B	two letters with a bar above, such as $\overline{AB}$
angle	an amount of a turn	A B C	the symbol $\angle$ followed by three letters, such as $\angle ABC$
plane	a flat region, it extends forever	A B C	any three letters, such as ABC, that represent points that are not on the same straight line

PRACTICE

1 Draw a point and label it E.

2 Draw a line segment and label it with points C and B.

3 Draw a line and label it with points V and W.

4 Draw an angle and label it with points D, F, and G.

5 The symbol $\angle$ is the symbol for a(n) _?_.

_____

6 The symbol $\leftrightarrow$ is the symbol for a(n) _?_.

_____

Use this illustration of a house to answer questions 7 and 8.

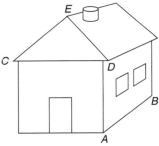

7 Which of these would lie in the same plane if the house were real?

A the chimney and the door
B the windows
C all of the above

8 The peaked roof is formed by two planes. Do the two planes meet in a point, in a plane, or in a line?

_____

Lines

Two lines that cross or that will cross are called **intersecting lines**.

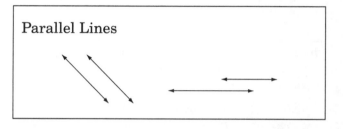

Intersecting Lines

Two lines that are always the same distance apart are called **parallel lines**.

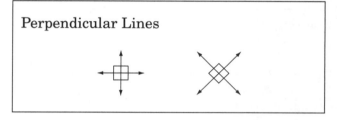

Parallel Lines

Two lines that form right angles when they meet are called **perpendicular lines**.

Perpendicular Lines

PRACTICE

Use this diagram to answer questions 1–3.

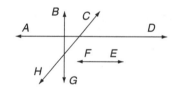

1   Which line is parallel to $\overleftrightarrow{AD}$? _____

2   Which two lines are perpendicular? _____

3   $\overleftrightarrow{BG}$ and $\overleftrightarrow{HC}$ are _?_ lines. _____

Use this map to answer questions 4–6.

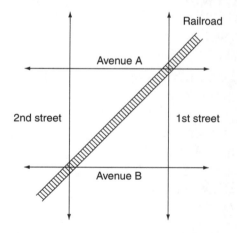

4   Which street(s) are parallel to 1st street?

5   Name two perpendicular streets.

6   In this map, railroad ties would be represented as _?_ .

Angles

The **size** of an angle refers to the opening between the sides of the angle. The point where the sides meet is called the **vertex** of the angle.

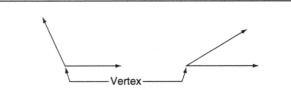

The angle at the left is larger than the angle at the right.

You can think of an angle as part of a circle. A complete circle is 360 degrees (or 360°). At the right, look at the angle made by a square corner with its vertex at the center of the circle. That angle represents $\frac{1}{4}$ of a circle, so its measurement is $\frac{1}{4}$ of 360° or 90°.

A full circle is 360°.
One-quarter of a circle is $\frac{360}{4} = 90°$.

One-third of a circle is $\frac{360}{3} = 120°$.

One-eight of a circle is $\frac{360}{8} = 45°$.

PRACTICE

In each pair below, circle the letter for the smaller angle.

1

2

3
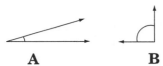

4
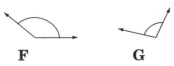

For each angle, circle the letter for the best estimate of its measurement. If you have a protractor, use it to find the actual measurement of the angle.

5   **A**   120°
      B   90°
      C   35°

6   **F**   100°
      G   150°
      H   50°

7   **A**   75°
      B   50°
      C   25°

8   **F**   80°
      G   130°
      H   90°

Types of Angles

An angle whose measurement is 90° is called a **square corner** or a **right angle.** An angle smaller than a right angle is called an **acute angle.** An angle larger than a right angle (but smaller than 180°) is called an **obtuse angle.**

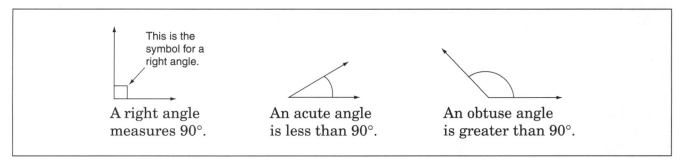

A right angle
measures 90°.

An acute angle
is less than 90°.

An obtuse angle
is greater than 90°.

PRACTICE

For each angle below, write whether it is acute, right, or obtuse. You may use the square corner of a piece of paper.

1

_____

2

_____

3

_____

4

_____

5

_____

6

_____

7

_____

8

_____

9

_____

10

_____

Use this diagram to answer questions 11 through 13.

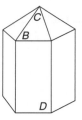

Base
A

11   Tell what type of angle is formed at each corner.

A   _____

B   _____

C   _____

D   _____

12   Each side panel in the gazebo has four corners. If you find the total of the measures of the angles formed by these four corners, the total is _?_.

_____

13   If you find the total of the six angles in the base, that total is _?_.

F   exactly 540°
G   more than 540°
H   less than 540°

Related Angles

You can think of a straight line as an angle whose measure is 180°. If you can put two angles together so their outer sides form a straight line, the angles are called **supplementary angles.** If you know the measure of one angle in a supplementary pair, you can find the measure of the other angle.

Angles that form a right angle when they are put together are called **complementary angles.** If you know the measure of one of two complementary angles, you can subtract it from 90° to find the measure of the other angle.

When two lines intersect, they form two pairs of equal angles. The equal angles, called **vertical angles,** are the angles that are opposite or across from each other.

Supplementary Angles	Complementary Angles	Vertical Angles
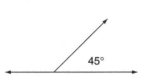		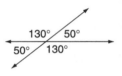
The sum of these two angles is 180°, so the measure of the unlabeled angle is 180° − 45° = 135°.	The measure of the unmarked angle is 90° − 30° = 60°.	

PRACTICE

Find the measure of the unmarked angle(s) for each of the following problems.

1

2

3

4
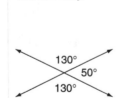

5

Use the figure below to answer questions 6 though 9.

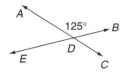

6   ∠*CDE* and ∠*EDA* are ___?___ .

   A   supplementary
   B   complementary
   C   vertical

7   What angle is a vertical angle to ∠*BDC?*

_____

8   What is the measure of ∠*BDC?*

_____

9   What is the measure of ∠*CDE?*

_____

The figure below is a rectangle. Each of its corners is a right angle. Use the rectangle to answer questions 10 through 12.

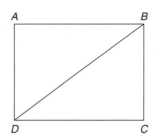

10   What is the measure of ∠*BCD?*

_____

11   Which of these angles is a supplement to ∠*BCD?*

   F   ∠*CDB*
   G   ∠*BDA*
   H   ∠*CDA*

12   What angle is a complement to ∠*CBD?*

_____

When a line intersects two parallel lines, all the angles have one of two measures, as shown in the diagrams below. Use the patterns in the measures of these angles to answer questions 13–16.

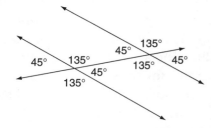

13   Which angle is equal to ∠*s?*

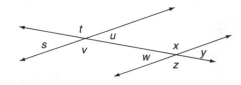

   A   ∠*v*
   B   ∠*t*
   C   ∠*w*
   D   ∠*x*

14   Which angle is equal to ∠*z?*

   F   ∠*y*
   G   ∠*u*
   H   ∠*v*
   J   ∠*s*

15   Which angle is a supplement to ∠*t?*

   A   ∠*v*
   B   ∠*w*
   C   ∠*x*
   D   ∠*z*

16   What angle is opposite ∠*s?*

_____

Related Angles

Polygons

A **polygon** is a figure that has straight sides. A polygon is named for the number of sides it has.

A triangle has 3 sides.
A quadrilateral has 4 sides.
A pentagon has 5 sides.
A hexagon has 6 sides.
An octagon has 8 sides.

Special Types of Plane (Flat) Figures

Name	Examples	Definition
regular polygon	△ ▢	a polygon with equal sides and equal angles
parallelogram	▭ ▱	a 4-sided figure with opposite sides that are parallel and equal
rectangle	▭ ▭	a parallelogram with four right angles. Opposite sides have the same length.

Name	Examples	Definition
square	⊞	a rectangle with four equal sides. Since it is a rectangle, it has four right angles.
trapezoid	⏢	a 4-sided figure with exactly one pair of parallel sides
rhombus	▱ ◇ ◻	a parallelogram with 4 equal sides

PRACTICE

This diagram shows the play areas in a neighborhood park. Use the diagram for questions 1–6.

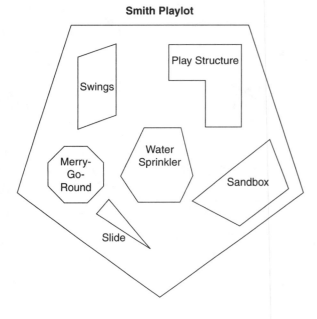

Smith Playlot

1   Which region is a parallelogram? _____

2   Which region is a hexagon? _____

3   The sandbox forms a(n) _?_. _____

4   The playlot as a whole forms a regular _?_. _____

5   Which region has the most sides? _____

6   There is only one figure inside the playlot that is a regular polygon. What play area is it and what figure does it form? _____

Types of Triangles

Triangles are a particularly interesting and important type of figure. For the sides, the sum of any two sides must be greater than the third side.

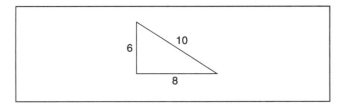

6 + 8 is greater than 10.
6 + 10 is greater than 8.
8 + 10 is greater than 6.

Here is another important thing to know about all triangles:

> If you add the measurements of all three angles in a triangle, the sum is always 180°.

Some triangles have three equal sides and three equal angles. Other triangles have two equal sides and two equal angles. Still other triangles have sides that are three different lengths and angles that are three different sizes.

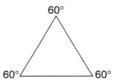

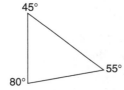

Equilateral triangle	**Isosceles triangle**	**Scalene triangle**
All sides are equal.	At least two sides are equal.	No two sides are equal.
All angles are equal.	At least two angles are equal.	No two angles are equal.
60° + 60° + 60° = 180°	70° + 70° + 40° = 180°	80° + 45° + 55° = 180°

PRACTICE

Use the classifications above to name each type of triangle. Then find out the measure of the unlabeled angle.

1   Type of triangle:

_____

Measure of the
unlabeled angle:

_____

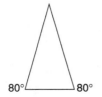

2   Type of triangle:

_____

Measure of the
unlabeled angle:

_____

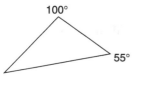

3   Type of triangle:

_____

Measure of the
unlabeled angle:

_____

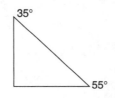

4   Type of triangle:

_____

Measure of the
unlabeled angle:

_____

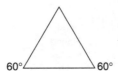

Right Triangles

A triangle that contains a **right angle** (an angle whose measure is 90°) is called a **right triangle.** The side opposite the right angle is called the **hypotenuse.** The other two sides of a right triangle are called the **legs.** The hypotenuse is always the longest side in a right triangle.

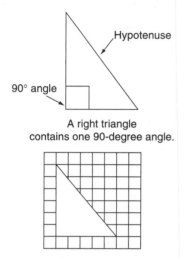

90° angle

Hypotenuse

A right triangle contains one 90-degree angle.

> ### The Pythagorean Theorem
>
> In a right triangle, the square of the hypotenuse is equal to the sum of the squares of the other two sides.

One leg of this triangle is 5 units long. The other leg is 6 units long.

$$\begin{aligned}(\text{hypotenuse})^2 &= 5^2 + 6^2 \\ &= 25 + 36 \\ &= 61 \\ \text{hypotenuse} &= \sqrt{61}\end{aligned}$$

PRACTICE

Name the right triangles in each window pane below. If there are no right triangles, write "None."

1

_____
right triangles

2

_____
right triangles

3

_____
right triangles

Find the length of the hypotenuse in each right triangle below. Give each answer as the square root of a whole number.

4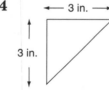

3 in.

3 in.

_____
hypotenuse

5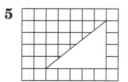

_____
hypotenuse

6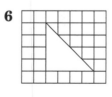

_____
hypotenuse

7 This diagram shows a pole leaning against a wall.

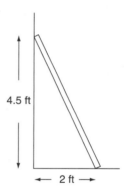

4.5 ft

2 ft

How long is the pole?

_____

Congruent Figures and Similar Figures

If two figures have the same size and the same shape, they are called **congruent** figures.	If two figures have the same shape, they are called **similar** figures. (Two similar figures *can be* the same size.)

 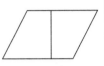

four congruent squares

two congruent triangles

two congruent trapezoids

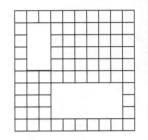

PRACTICE

Use this diagram to answer questions 1–3.

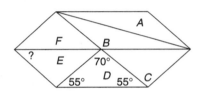

1 Which triangle is congruent to triangle *A*?

_____

2 Which triangle is similar to triangle *F*?

_____

3 Triangles *E* and *D* are congruent. What is the measure of the angle marked with a question mark?

_____

4 These two figures are similar.

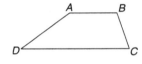

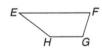

Which side in figure *EFGH* **corresponds to** (is in the same position as) side *AB* in figure *ABCD?*

_____

5 Are these two triangles similar?

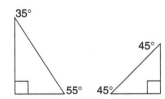

_____

The triangles below are similar. Use them to do questions 6–8.

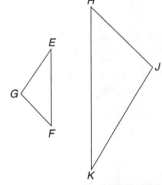

6 Which side of triangle *HJK* corresponds to side *EF* in triangle *EFG?*

_____

7 Which angle in triangle *HJK* is equal to ∠*FGE?*

_____

8 Which side of triangle *HJK* corresponds to side *GE* in triangle *EFG?*

_____

Solving Problems with Similar Triangles

When you are working with similar triangles, you can use the following facts to find distances.

The corresponding angles in similar figures have the same measure.	The corresponding sides of similar figures form proportions.

"Corresponding sides" are sides in similar figures that are in the same position. In triangles, corresponding sides are always opposite equal angles. We know that the triangles on the right are similar because they are made by intersecting lines between two parallel lines. Therefore, we can deduce the following:

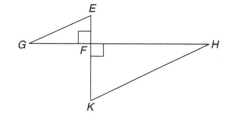

Side *KH* must correspond to side *EG* since they are both opposite right angles.
We can see that angles *FGE* and *FHK* are the smallest angles in each triangle, so they must be equal. That means the sides opposite them must correspond: side *EF* corresponds to side *KF*.
The only remaining sides are *HF* and *GF*, so they must correspond.

Then we can set up the following proportions. Notice that the numerators always get sides from the same triangle. So do the denominators.

$$\frac{KH}{EG} = \frac{KF}{EF} = \frac{HF}{GF}$$

To use this information to solve problems, put the lengths of known sides into ratios. Then solve proportions to find the lengths of unknown sides.

$$\frac{KF}{EF} = \frac{HF}{GF} \text{, so}$$

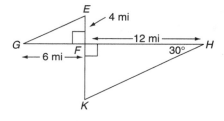

$$\frac{KF}{4} = \frac{12}{6}$$
$$6 \times KF = 4 \times 12$$
$$6 \times KF = 48$$
$$KF = \frac{48}{6} = 8$$

KF is 8 miles.

PRACTICE

Use your knowledge of triangles to solve problems 1 and 2.

1   Side *EG* in the figure above is 7 miles long. How long is side *KH* in the similar triangle? _____

2   Which angle in the figure above has the same measure as ∠*HKF*? _____

In the diagram below, lines *UW* and *TY* are parallel. Study the diagram. Then use it to answer questions 3 through 10.

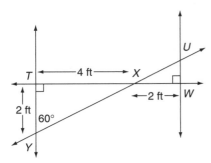

3 What is the measure of ∠*UWX*? _____

4 Which side forms the hypotenuse in right triangle *XTY*? _____

5 What is the measure of ∠*TXY*? _____

6 Using the Pythagorean Theorem, you can deduce that the length *XY* is the square root of __?__ . _____

7 What is the measure of ∠*XUW*? _____

8 Which side of triangle *UXW* corresponds to side *XY* in triangle *TXY*? _____

9 What is the length of side *UW*? _____

10 Which of these angles is a supplement to ∠*UXW*?

    A ∠*UXT*
    B ∠*UWX*
    C ∠*TXY*

The drawing below shows two similar triangles created by the rays of the sun and shadows cast by a tree and a 6-foot-tall person. Study the drawing. Then use it to answer questions 11 through 14.

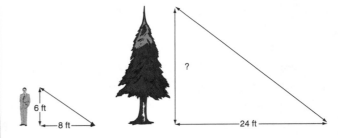

11 Which type of triangle has been formed in both cases?

    F isosceles
    G equilateral
    H right

12 Using the Pythagorean Theorem, you can deduce that the distance from the person's head to the far end of the person's shadow is __?__ feet

_____

13 In these triangles, the person's height corresponds to __?__ .

    A the length of the tree's shadow
    B the sun's rays passing from the tree to the ground
    C the tree's height

14 How tall is the tree?

_____

Solving Problems with Similar Triangles

Geometry Skills Practice

Read the passage and study the map below. Then do Numbers 1 through 6.

This map shows the Granville family farm and all the roads nearby. Highway 32 and Rondell Road are parallel. The farmhouse is at the intersection of Lacy Lane and Flint Road. That intersection is 1 mile from Rondell Road via Flint Road, and it is 3 miles from Highway 32 via Flint Road.

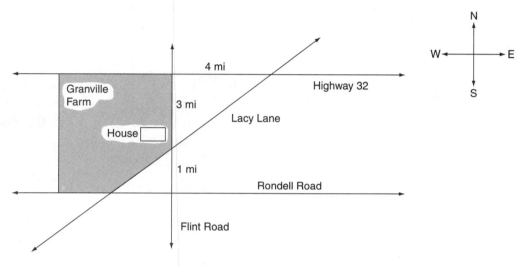

1 Which of these roads are perpendicular?

    A Highway 32 and Rondell Road
    B Lacy Lane and Flint Road
    C Rondell Road and Flint Road
    D Rondell Road and Lacy Lane

2 What type of angle is formed by the corner of the Granville property where the farmhouse sits?

    F an obtuse angle
    G an acute angle
    H a right angle
    J a supplementary angle

3 What kind of polygon do the property lines of the Granville farm create?

A   a parallelogram
B   a quadrilateral
C   a pentagon
D   an octagon

4 The north and west borders of the Granville farm are both 4 miles long. Which of these is the best estimate of the farm's total area?

F   22 square miles
G   20 square miles
H   15 square miles
J $13\frac{1}{3}$ square miles

5 At the intersection of Highway 32 and Flint Road, you can reach Lacy Lane by driving 4 miles east or by driving 3 miles south.

How far is the farmhouse from Highway 32 via Lacy Lane?

A   49 miles
B   25 miles
C   7 miles
D   5 miles

6 The two triangles created by the roads on this map are similar. What is the distance between Flint Road and Lacy Lane when you are driving along Rondell Road?

F $\frac{3}{4}$ mile
G $1\frac{1}{3}$ miles
H   3 miles
J   4 miles

Skills Inventory Post-Test

Part A: Computation

Circle the letter for the correct answer to each problem.

1

$9.802 + 0.08 = $ ___

- **A** 10.602
- **B** 9.81
- **C** 9.882
- **D** 9.722
- **E** None of these

2

$59.2 - 5.8 = $ _____

- **F** 1.2
- **G** 54.4
- **H** 53.4
- **J** 66
- **K** None of these

3

What is 60 percent of 115?

- **A** 69
- **B** 52.2
- **C** 19.2
- **D** 73
- **E** None of these

4

$16 \times (-4) = $ _____

- **F** 4
- **G** -4
- **H** 64
- **J** -64
- **K** None of these

5

$\begin{array}{r} \frac{2}{3} \\ -\frac{1}{6} \\ \hline \end{array}$

- **A** $\frac{1}{6}$
- **B** $\frac{1}{2}$
- **C** $\frac{1}{3}$
- **D** $\frac{1}{9}$
- **E** None of these

6

$0.072 \div 6 = $ _____

- **F** 12
- **G** 0.12
- **H** 0.012
- **J** 0.0012
- **K** None of these

7

$5\overline{)1.75}$

- **A** 0.31
- **B** 3.1
- **C** 0.41
- **D** 3.5
- **E** None of these

8

$3\frac{1}{5} \times \frac{2}{3} = $ _____

- **F** $2\frac{2}{15}$
- **G** $3\frac{2}{15}$
- **H** $\frac{2}{3}$
- **J** $\frac{8}{15}$
- **K** None of these

9

$-15 + (-2) = $ _____

- **A** 13
- **B** -13
- **C** -17
- **D** 17
- **E** None of these

10

$3^2 \times 10^4 = $ _____

- **F** 30^6
- **G** 90,000
- **H** 300,000
- **J** 30^8
- **K** None of these

11 $[14 - (9 - 2)]^2 = $ ___

- **A** 9
- **B** 49
- **C** -35
- **D** -63
- **E** None of these

12

$\begin{array}{r} 90.15 \\ \times \quad 6 \\ \hline \end{array}$

- **F** 540.9
- **G** 540.6
- **H** 54.9
- **J** 5.409
- **K** None of these

13

$3x(x^2 + xy^2) = $ ___

A $4x^2 + 3x^2y^2$
B $3x^3 + 3xy^2$
C $3x^3 + 3x^2\,y^2$
D $3x^4y^2$
E   None of these

14

What is 125 percent of 42?

F   300
G   52.5
H   33.6
J   525
K   None of these

15

What percent of $15.00 is 60 cents?

A   4%
B   25%
C   2.5%
D   40%
E   None of these

16

$9\frac{1}{4} - 2\frac{1}{2} = $ ___

F   7
G $6\frac{3}{4}$
H $6\frac{1}{2}$
J $13\frac{1}{2}$
K   None of these

17

$\frac{-15}{-5} = $ ___

A   −5
B   5
C   3
D   −3
E   None of these

18

320% of ☐ = 96

F   307.2
G $33\frac{1}{3}$
H   30
J   300
K   None of these

19 $2x + x(3 + xy + y^2) = $ ___

A $5x^2 + x^2y + xy^2$
B $6x + x^2y + xy^2$
C $5x + x^3y^3$
D $5x + x^2y + xy^2$
E   None of these

20 $xy + x + x = $ ___

F $3xy$
G $2xy$
H x^3y
J $xy + 2x$
K   None of these

21

30% of ☐ = 21

A   70
B   63
C   14.2
D   76
E   None of these

22

$\frac{1}{2} + \frac{3}{4} + \frac{1}{12} = $ ___

F $1\frac{1}{4}$
G $\frac{1}{4}$
H $\frac{1}{3}$
J $1\frac{1}{3}$
K   None of these

23

$5\frac{1}{4} \div 2\frac{1}{3} = $ ___

A $10\frac{1}{12}$
B $2\frac{1}{4}$
C $12\frac{1}{4}$
D $5\frac{1}{4}$
E   None of these

24

$-2 - (-7) = $ ___

F   5
G   −5
H   −9
J   9
K   None of these

25

$6 + (-8) = $ ___

A   −14
B   14
C   2
D   −2
E   None of these

Part B: Applied Mathematics

1 You are estimating by rounding to the nearest whole number. What numbers should you use to estimate 8.02 times 1.87?

    A 8 and 1
    B 8 and 2
    C 8.0 and 1.9
    D 8.0 and 1.8

2 If you change the 3 in 234,679 to a 4, it increases the value of the number by __?__ .

    F one
    G one thousand
    H ten thousand
    J one hundred thousand

3 Manuel arrives at the public library at 8:54. He studies until 10:09. How many hours did he study?

    A $\frac{1}{4}$ hr   **C** $1\frac{1}{4}$ hr

    B $1\frac{1}{2}$ hr   **D** $1\frac{1}{3}$ hr

4 For which of these situations could you use an estimate?

    F writing checks
    G reporting finishing times in a race
    H finding the distance between two towns
    J making out time cards

5 Which of these statements is true?

    A $-4 > -2$ **C** $-4 > 0$
    B $-4 > 3$ **D** $-4 > -5$

6 Between what two whole numbers is $\sqrt{15}$?

    F 14 and 15 **H** 3 and 4
    G 2 and 3   **J** 4 and 5

7 Which of these numbers, when added to 0.09, produces a sum greater than zero?

    A 0.1   **C** 0.12
    B 0.02   **D** 0.92

8 Which of these fractions is greater than $\frac{1}{2}$?

    F $\frac{3}{7}$ **H** $\frac{5}{6}$

    G $\frac{2}{5}$ **J** $\frac{4}{9}$

9 On a buying trip, Kurt buys 50 sweaters for $9.50 each. He brings them home and sells them for $50.00 each at a craft fair. Which of these number sentences can Kurt use to figure out how much money he made?

    A $50(\$50.00) - 50(\$9.50) = \square$
    B $50(\$50.00) \times 50(\$9.50) = \square$
    C $50(\$50.00) = \square$
    D $50(\$50.00) - \$9.50 = \square$

10 One serving of ice cream is $\frac{1}{2}$ cup. How many servings of ice cream are there in $\frac{1}{2}$ gallon?

    F 8   **H** 16
    G 5   **J** 32

11 Which of these equations represents the relationship that two times the difference between 10 and a number is 40?

    A $2n - 10 = 40$
    B $2(10 - n) = 40$
    C $2(10) - n = 40$
    D $2(40) = 10 - n$

12 Which of these numbers has a value between -2 and -6?

    F -7 **H** -1
    G 0   **J** -3

This table gives parents an idea of how much money they can expect to spend raising their children. It shows the average amount that experts predict middle-class parents will spend each year per child. Use the table for Numbers 13–17.

The Yearly Costs of Raising a Child Born in 1996*

Child's Age	Expenses	Child's Age	Expenses
under 1	$7,880	10	13,450
1	8,270	11	14,150
2	8,700	12	16,220
3	9,380	13	17,070
4	9,870	14	17,950
5	10,390	15	19,170
6	11,020	16	20,160
7	11,590	17	21,210
8	12,200	**Total**	241,440
9	12,780		

Source: Family Economics Research Group, U.S. Department of Agriculture

13 According to this table, what fraction of the total cost of raising a child is spent when the child is 17?

A about $\frac{1}{12}$ C about $\frac{1}{3}$

B about $\frac{1}{9}$ D about $\frac{1}{5}$

14 How much does the yearly cost of raising a child increase between the child's first and eighteenth years? (Estimate, using numbers rounded to the thousands place.)

F 333$\frac{1}{3}$% H 192%

G 162% J 163%

15 Sending a child to college increases the cost of raising the child by about 50%. A middle-class parent who sends a child to college can expect to spend about how much on the child altogether?

A $200,000 C $400,000
B $1,200,000 D $360,000

16 The greatest increase in the cost of raising a child occurs when the child is age __?__.

F 1 H 12
G 8 J 17

17 Two parents decide to make a graph showing how much they can expect to pay each year of their child's life. Which of these graphs shows the general relationship between the child's age and the yearly expense of raising the child?

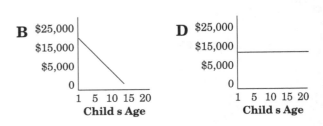

This graph shows a state's income in 1996. Study the graph. Then use it to do Numbers 18 through 21.

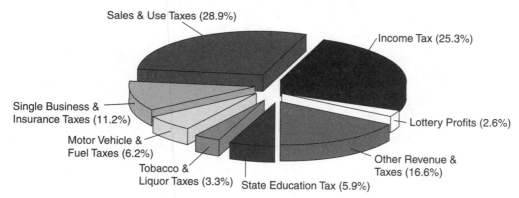

Total of State Revenues: $21,000,000,000

Sales & Use Taxes (28.9%)

Income Tax (25.3%)

Single Business & Insurance Taxes (11.2%)

Motor Vehicle & Fuel Taxes (6.2%)

Tobacco & Liquor Taxes (3.3%)

State Education Tax (5.9%)

Lottery Profits (2.6%)

Other Revenue & Taxes (16.6%)

18 What fraction of the state's income came from income taxes?

F about $\frac{1}{2}$

G about $\frac{1}{3}$

H about $\frac{1}{4}$

J about $\frac{1}{5}$

19 How much were state lottery profits in 1996?

A $5,460 million
B $5.46 million
C $54.6 million
D $546 million

20 What percentage of the state's income came from a combination of lottery profits, motor vehicle and fuel taxes, and single business and insurance taxes?

F 15%
G 20%
H 25%
J 30%

21 How much more did the state make through sales and use taxes than it made through income taxes?

A $3.6 million
B $756 million
C $606 million
D $360 million

The table below shows some of the legal grounds for an "absolute" (or one-sided) divorce in different western states. It also shows how long you must live in each state before a divorce can be granted. Study the table. Then do Numbers 22 through 26.

Grounds for Absolute Divorce in the Western States

	Adultery	Cruelty	Desertion	Alcoholism	Non-support	Insanity	Bigamy	Felony Conviction	Drug Addiction	Fraud, Force, or Duress	Required Residency
Alaska	yes	yes	1 yr	1 yr	no	18 mo	yes[A]	yes	yes	yes[A]	none
Arizona	no	no	no	no	no	no	no	no	no	no	90 days
California	no	no	no	no	no	yes	yes[A]	no	no	yes[A]	6 mo*
Colorado	no	no	no	no	no	no	yes[A]	no	no	yes[A]	90 days
Idaho	yes	yes	yes	no	no	3 yr	yes[A]	yes	no	yes[A]	6 wk
Montana	no	no	no	no	no	no	yes[A]	no	no	yes[A]	90 days
Nevada	no	no	no	no	no	2 yr	yes[A]	no	no	yes[A]	6 wk
New Mexico	yes	yes	yes	no	no	no	no	no	no	no	6 mo
Oregon	no	no	no	no	no	no	no	no	yes	yes[A]	6 mo*
Utah	yes	yes	1 yr	yes	yes	yes*	yes[A]	yes	no	no	3 mo*
Washington	no	no	no	no	no	no	yes[A]	no	no	yes[A]	none
Wyoming	no	no	no	no	no	2 yr	yes[A]	no	no	yes[A]	2 mo*

[A]Grounds for annulment
*Not allowed in all cases

22 Which of the reasons listed is recognized by the fewest western states?

F adultery **H** bigamy
G cruelty **J** non-support

23 For a divorce based on adultery, what is the ratio of western states that allow absolute divorce on those grounds to western states that do not allow divorce on those grounds?

A 3 : 1 **C** 2 : 1
B 1 : 3 **D** 1 : 2

24 Montana's required period of residency is about $\frac{1}{2}$ as long as the time period required by ___?___ .

F Wyoming **H** New Mexico
G Utah **J** Nevada

25 According to this table, which western state makes it hardest to get an absolute divorce?

A Arizona
B California
C New Mexico
D Utah

26 Which of these states takes a different stance toward bigamy than it does toward fraud, force, or duress?

F Alaska
G Idaho
H New Mexico
J Utah

A rock climber 6 feet tall needs to figure the height of this cliff. He puts a pan of water on level ground at point *C* and backs up until he sees the top of the cliff reflected in the water. He knows that creates two similar triangles as shown in the diagram. Study the diagram. Then do Numbers 27 through 31.

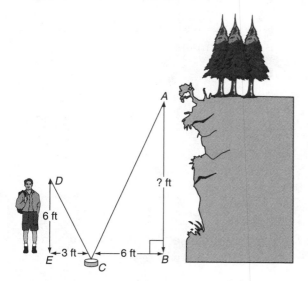

27 What is the distance from point *C* to point *D*?

A 3 ft

B $\sqrt{45}$ feet

C 6 feet

D $\sqrt{18}$ feet

28 What angle is congruent to ∠*EDC*?

F ∠*DCA*

G ∠*ECD*

H ∠*BCA*

J ∠*CAB*

29 How tall is the cliff?

A 10 feet

B 12 feet

C 16 feet

D 18 feet

30 This method will not work if the climber ___?___ .

F looks at the pan with one eye

G leans forward as he stands

H uses a deep pan of water

J stands on one foot

31 The climber decides to make these measurements in meters as well as feet. What is the climber's height in meters? (1 meter ≈ 3.3 feet)

A about 1.8 meters

B about 19.8 meters

C about 0.55 meter

D about 2.2 meters

32 Which group of integers is in order from least to greatest?

F 0, −11, −9, 2, 3, 15

G 0, 2, 3, −9, −11, 15

H −11, −9, 0, 2, 3, 15

J −9, −11, 0, 2, 3, 15

Read this passage and study the diagram. Then do Numbers 33 through 36.

A shopping mall is planning to build a bench around a circular planter. The planter will be 6 feet in diameter and $2\frac{1}{2}$ feet deep. The bench will be 7 feet by 7 feet with angled corners. A diagram of the bench and planter are given below. Use the following when necessary:

$\pi = 3.14$

volume of a cylinder $= \pi r^2 h$

area of a triangle $= \frac{1}{2}bh$

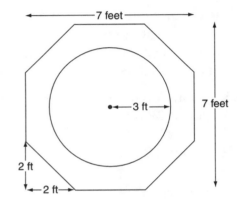

33 What type of polygon does the bench form?

   A a pentagon

   B a hexagon

   C a trapezoid

   D an octagon

34 What area will the bench and the planter cover?

   F 49 square feet

   G 45 square feet

   H 41 square feet

   J 33 square feet

35 If the soil in the planter is $2\frac{1}{2}$ feet deep, how much soil is in the planter?

   A about 282 cubic feet

   B about 71 cubic feet

   C about 24 cubic feet

   D about 122 cubic feet

36 How wide will the bench be at its most narrow width?

   F 1 foot

   G $\frac{1}{2}$ foot

   H 2 feet

   J $1\frac{1}{2}$ feet

37 Which of these numbers has a square root that is a whole number?

   A 13

   B 12

   C 16

   D 30

38 What value goes in the box to make the number sentence true?

$$5.24 \times \square = 52{,}400$$

   F 10^2

   G 10^3

   H 10^4

   J 10^6

Read this advertisement. Then do Numbers 39 through 44.

Save from 50% to 80% on calls to these countries!

Charges for a 2-min call from the U.S.

	Clear Tone*	**"The other guys"**
Australia	$0.60	$3.00
Bahamas	0.60	2.40
Brazil	1.00	3.50
China	1.75	5.25
Cuba	1.70	3.40
France	0.60	2.40
Greece	1.00	4.00
Italy	0.80	3.20
Japan	0.70	2.80
Philippines	1.30	3.90
Poland	1.00	3.00
Sweden	0.70	2.10
Britain	0.40	2.00

***$6.25 monthly service charge**

Just 15.9 cents for the first two minutes of a call within the U.S.
28 cents for each additional minute.

---

39 For which country is Clear Tone's charge 50% of the "other guy's" charge?

   A Australia
   B Brazil
   C Cuba
   D Greece

40 For which two countries does Clear Tone offer the biggest percent savings (80% off)?

   F Australia and Britain
   G Australia and Bahamas
   H Bahamas and France
   J Australia and China

41 Which of these equations could you use to calculate Clear Tone's charge (c) in cents for a call within the U.S? (Let m stand for the length of the call in minutes.)

   A $c = 15.9m$
   B $c = 15.9(m - 2)$
   C $c = 15.9 + 28m$
   D $c = 15.9 + 28(m - 2)$

42 Clear Tone charges a monthly service fee of $6.25, while the "other guys" have no monthly service fee. If you only used your phone to call Greece, you would save money if you used the "other guys" to make __?__ .

   F fewer than three 2-minute calls a month
   G three or four 2-minute calls a month
   H four or five 2-minute calls a month
   J more than five 2-minute calls a month

43 Last month, Ming used Clear Tone to make four 2-minute calls to China and two 2-minute calls to Britain. How much did he spend on international calls?

   A $7.00   **C** $14.05
   B $2.15   **D** $17.40

44 Flora buys a package with Clear Tone that gives her unlimited calls within her area code for $22.00 a month, including the service fee. The only long-distance call she makes is a call to Poland once a week. If she keeps those calls down to 2-minutes each, how much will her yearly phone bill be?

   F $264.00
   G $74.00
   H $316.00
   J $176.00

Study this bar graph and this table. Then do Numbers 45 through 50.

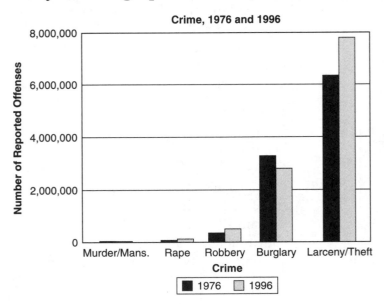

Crime, 1976 and 1996

Percent Change in the Number of Offenses, 1976 to 1996	
Murder/Manslaughter	−2.2
Rape	+5.1
Robbery	+3.7
Burglary	−22.7
Larceny/Theft	+5.3

45 Which crime rate changed the most between 1976 and 1996?

   A rape   C burglary
   B robbery   D larceny/theft

46 Which of these is the best estimate of the number of burglaries in 1996?

   F 3,500,000   H 2,000,000
   G 2,500,000   J 1,500,000

47 Which of these claims is supported by the information in these graphs?

   A The war against drugs has decreased crime.
   B The rapid construction of new prisons during the 1990s has decreased crime.
   C The rapid rise in use of home security systems has discouraged burglars.
   D The rapid rise in the use of car alarms has discouraged car thieves.

48 Which of these is the best estimate of the average rise in larceny thefts each year from 1976 to 1996?

   F 170,000   H 70,000
   G 85,000   J 35,000

49 A formula can be created to describe the number of rapes per thousand people in a year. Using R to represent the total number of rapes that year and using N to represent the number of rapes per 1,000 people, the formula is $N = \frac{R}{P}$. What is P?

   A the number of rapes in the previous year
   B the number of rapists in the U.S.
   C the population of the U.S. in thousands
   D the number of women in the U.S.

50 From 1996 to 1997, the percent change in the number of rapes per 1,000 people was −3.5. What does this suggest about the proportion of people affected by rape in 1996 and 1997?

   F Any one person was less likely to be raped in 1997 than in 1996.
   G There were fewer rapes committed in 1997 than in 1996.
   H There were fewer rapists in 1997, but they committed more crimes each.
   J There were more rapists in 1997, but most of them were imprisoned and could not commit any crimes.

Skills Inventory Post-Test Evaluation Chart

Use the keys to check your answers on the Post-Test. The Evaluation Chart shows where you can turn in the book to find help with the problems you missed.

Keys

Part A

1	C	14	G
2	H	15	A
3	F	16	G
4	J	17	C
5	B	18	H
6	H	19	D
7	E	20	J
8	F	21	A
9	C	22	J
10	G	23	B
11	B	24	F
12	F	25	D
13	C		

Evaluation Charts

Part A

Problem Numbers	Skill Areas	Practice Pages
1, 2, 6, 7, 12	Decimals	20–31
5, 8, 16, 22, 23	Fractions	32–46
4, 9, 17, 24, 25	Integers	47–53
3, 14, 15, 18, 21	Ratios/Proportions/Percents	54–64
10, 11, 13, 19, 20	Algebraic Operations	80, 83, 88, 93–97, 100–101

Part B

1	B	26	J
2	H	27	B
3	C	28	J
4	H	29	B
5	D	30	G
6	H	31	A
7	D	32	H
8	H	33	D
9	A	34	H
10	H	35	B
11	B	36	G
12	J	37	C
13	A	38	H
14	G	39	C
15	D	40	F
16	H	41	D
17	A	42	F
18	H	43	C
19	D	44	H
20	G	45	C
21	B	46	G
22	J	47	C
23	D	48	G
24	H	49	C
25	A	50	F

Part B

Problem Numbers	Skill Areas	Practice Pages
2, 5, 6, 7, 8, 12, 18, 23, 32, 39	Numeration/Number Theory	1, 20–21, 32, 47, 78, 83, 88
15, 17, 22, 25, 26, 40, 45, 47, 50	Data Interpretation	65–77
9, 11, 37, 38, 41, 42, 49	Pre-Algebra/Algebra	78–85, 86–104
3, 10, 24, 31, 34, 35	Measurement	105–121
27, 28, 29, 30, 33	Geometry	122–136
14, 19, 20, 21, 43, 44	Computation in Context	11–14, 29, 44, 51, 62, 70, 102, 119
13, 14, 16, 46, 48	Estimation/Rounding	15–17

Answer Key

Pages 1–2, Place Value
1. hundred thousands, thousands, hundreds
2. millions, ten-thousand, thousands
3. hundreds
4. hundred-thousands
5. 5
6. 7
7. twenty
8. seventy thousand
9. six hundred thousand
10. four hundred
11. eighty thousand
12. three hundred two
13. one thousand, forty
14. three thousand, one hundred
15. one million, three hundred fifty thousand
16. 12,040
17. 59,006
18. 312,000
19. 1,102
20. 10,500,000
21. one thousand
22. two hundreds
23. It decreases by two hundred thousand.
24. 7,100
25. 170,000
26. 100,000
27. 23,000
28. 10,605
29. 7,650
30. 14,679

Page 3, Review of Whole Number Addition
1. 91
2. 472
3. 1,480
4. 123 gal
5. 433 liters
6. 5,541
7. 12,725
8. 258 lb
9. 223
10. 1,380
11. 3,815 mi

Pages 4–5, Review of Whole Number Subtraction
1. 42 mi
2. 21
3. 790
4. 23 in.
5. 132
6. 491
7. 67
8. 1,212 ft
9. 15
10. 287
11. 78 kilometers
12. 573
13. 167 grams
14. 2,755
15. 2,791
16. 9,701
17. 73 dollars
18. 351
19. 61
20. 173
21. 4,951

Pages 6–7, Review of Whole Number Multiplication
1. 390
2. 600
3. 1,250 kilometers
4. 147 feet
5. 1,818
6. 2,642
7. 340
8. 18,300
9. 4,080
10. 3,344
11. 432 inches
12. 4,750
13. 88,998
14. 2,500
15. 6,360
16. 840 feet
17. 17,992
18. 13,899
19. 76,032

Pages 8–9, Review of Whole Number Division
1. 213
2. 4 r 1
3. 61 miles
4. 51 gallons
5. 5 R 4
6. 804
7. 7 r 2
8. 102
9. 10,204
10. 10,002
11. 357
12. 230 hours
13. 167
14. 88 liters
15. 53
16. 2,286
17. 43 meters
18. 564
19. 3,394

Page 10, Dividing by a Two-Digit Divisor
1. 11 r 3
2. 25 r 45
3. 202 r 8
4. 231 r 11
5. 23 r 3
6. 11 r 16
7. 22 r 19
8. 270
9. 211 r 28

Pages 11–12, Solving Word Problems: The First Steps
1. Subtract.
2. Multiply.
3. Add.
4. Divide.
5. Multiply.
6. Add.
7. A
8. H
9. A
10. G
11. C
12. G
13. C

Pages 13–14, Solving Word Problems: The Next Steps

1. $3\frac{1}{2}$, 5, and 6 should be circled.
2. $250.00, 2, and 3 days should be circled.
3. 6 girls, 6 ounces, and $0.25 per ounce should be circled.
4. 35 feet and 2 feet should be circled.
5. how much heating oil costs per gallon
6. how many miles she can drive per gallon
7. Multiply 25 by 15. Answer: 375
8. Add 16,420 and 1,567. Answer: 17,987
9. Subtract 95 from 1925. Answer: 1830
10. Divide 120 by 12. Answer: 10 ounces
11. Multiply $2,500 by 3. Multiply $3,000 by 9. Add the two products. Answer: $34,500
12. Subtract 84 dollars from 300 dollars. Divide the difference by 18. Answer: 12 dollars
13. Add 25 and 65. Then divide the sum by 45. Answer: 2 hours

Page 15, Estimation

1. circled
2. not circled
3. circled
4. circled
5. not circled
6. circled
7. not circled
8. C
9. F
10. B
11. C
12. G

Page 16, Rounding

1. 100
2. 400
3. 160
4. $79.40
5. $110
6. 6,000
7. 700
8. 1,700
9. 2,000
10. 12,000
11. $45.00
12. $140.00
13. $60.00
14. 56,100
15. 1,350
16. 2,426,000
17. 1,000
18. $18.00
19. $11.40
20. $110.00

Page 17, Rounding To Solve Word Problems

1. 130 miles + 250 miles = ☐
2. 210 miles − 160 miles = ☐
3. 100 miles + 110 miles = ☐
4. 140 miles × 2 × 3 = ☐
5. 490 miles
6. 8 gallons
7. 40 miles

Pages 18–19, Whole Numbers Skills Practice

1. A
2. H
3. B
4. H
5. A
6. F
7. B
8. H
9. B
10. H
11. C
12. G
13. B
14. H
15. C

Page 20, Decimal Place Values

1. thousandths
2. ten-thousandths
3. hundred-thousandths
4. hundredth
5. thousandths
6. three thousandths
7. 0.65
8. 0.403
9. 0.6
10. 5.106

Page 21, Comparing Decimal Numbers

1. 100
2. 10
3. 10
4. 1
5. $\frac{3}{100}$
6. $\frac{15}{100}$
7. $\frac{34}{1000}$
8. $\frac{903}{10000}$
9. 0.37
10. 0.09
11. 0.015
12. 0.0003
13. 1
14. 1.013
15. 12.059
16. 0.0607
17. 0.16
18. 0.5
19. 1.135
20. 7.429
21. 0.06, 0.6, 6
22. 0.01723, 0.018, 0.22, 10.28
23. 0.081, 0.091, 0.891

Page 22, Adding Decimals

1. 47.75 m
2. 0.020 or 0.02

3. 5.3329 million (or 5,332,900)
4. 0.2209
5. 0.205 g
6. 2.018
7. $0.75
8. $19.09
9. 137.13011
10. 63.02
11. 1.409
12. 0.6 or 0.60

Page 23, Subtracting Decimals

1. 14.041
2. 0.127
3. 21.375
4. 0.072
5. 0.8191
6. 4.943
7. 6.11 miles
8. 4.1915 grams
9. $24.97
10. $0.68 (or 68 cents)
11. 26.3991
12. 89.992
13. $9.5 million (or $9,500,000.00)
14. 11.46 mph

Pages 24–25, Multiplying Decimals

1. 0.006
2. 0.0036
3. 0.0104
4. 3.35
5. 20.048
6. 4.48
7. 0.000075
8. 0.0025
9. 46.789
10. 0.6799
11. 34.5
12. 0.5156
13. left
14. 3, left
15. 0.012
16. 1.26
17. 53
18. 1.35
19. 0.0055

20. 6.72
21. 1.5
22. $33.27
23. $61.20
24. $22.88
25. $41.52
26. $34.40
27. $6.30
28. $56.00
29. $2.14
30. $4.68
31. $275.00
32. 59.4 million

Page 26, Dividing a Decimal by a Whole Number

1. 0.0205
2. 3.05
3. 6.13
4. $5.08
5. $30.23
6. 0.562
7. 0.3
8. 0.25
9. 0.023
10. 0.031
11. 2.041
12. $1.90

Page 27, Dividing a Decimal by a Decimal

1. 44
2. 600
3. 11.2
4. 863
5. 12,700
6. 31,000
7. 20
8. 21.17
9. 70
10. 2,100
11. $4.10
12. 20
13. 23

Page 28, Adding Zeros at the Right of a Decimal

1. 1.5
2. 1.75

3. 11.67 (or $11\frac{2}{3}$)
4. 5.22
5. 3.55
6. 5.71
7. 4.09
8. 14.83
9. 0.33 (or $\frac{1}{3}$)

Page 29, Solving Mixed Word Problems

1. $17.37
2. $200.00
3. $4.10
4. 9 feet
5. 530.4 miles
6. 347.4 miles
7. $8.69
8. $3.59
9. $46.00

Pages 30–31, Decimals Skills Practice

1. A
2. H
3. D
4. F
5. D
6. J
7. A
8. F
9. C
10. G
11. A
12. J
13. B
14. K
15. B
16. J
17. C
18. J
19. B

Page 32, Numerators and Denominators

1. $\frac{646392}{2213582}$

2. $\frac{364512}{2213582}$

3. $\frac{140415}{224097}$

4. $\dfrac{646392}{1567190}$

5. $\dfrac{\$107800}{\$250000}$

6. $\dfrac{\$108000}{\$250000}$

7. $\dfrac{\$45000}{\$63000}$

8. $\dfrac{\$34200}{\$250000}$

Page 33, Finding a Fraction of a Number

1. 75
2. 150
3. 49
4. 315
5. 105
6. 105
7. 27.5
8. 1,312
9. $44.00
10. 120 pounds
11. 0.5
12. 0.25
13. 0.2
14. 0.75
15. 0.333...
16. 0.666...

Page 34, Reducing a Fraction to Simplest Terms

1. $\dfrac{1}{2}$
2. $\dfrac{1}{3}$
3. $\dfrac{1}{3}$
4. $\dfrac{7}{10}$
5. $\dfrac{1}{3}$
6. $\dfrac{3}{5}$
7. $\dfrac{11}{35}$
8. $\dfrac{13}{29}$
9. $\dfrac{1}{50}$
10. $\dfrac{11}{30}$

11. $\dfrac{1}{6}$
12. $\dfrac{11}{32}$
13. $\dfrac{1}{8}$
14. $\dfrac{2}{5}$
15. $\dfrac{1}{2}$
16. $\dfrac{1}{5}$
17. $\dfrac{1}{5}$
18. $\dfrac{1}{40}$
19. $\dfrac{1}{20}$

Page 35, Fractions Equal To and Fractions Greater Than 1

1. <
2. >
3. >
4. =
5. <
6. >
7. 5
8. 1
9. $1\dfrac{1}{2}$
10. 4
11. $1\dfrac{3}{7}$
12. $2\dfrac{1}{3}$
13. 3
14. $3\dfrac{3}{4}$
15. 2
16. $2\dfrac{1}{8}$
17. $2\dfrac{3}{4}$
18. 1
19. $1\dfrac{1}{2}$
20. 1
21. 5

Pages 36–37, Adding and Subtracting Like Fractions

1. 1
2. $1\dfrac{1}{5}$
3. $\dfrac{1}{8}$ cup
4. $\dfrac{1}{6}$
5. $5\dfrac{1}{2}$ m
6. $\dfrac{2}{3}$
7. $\dfrac{1}{2}$
8. $5\dfrac{2}{3}$
9. $1\dfrac{1}{3}$
10. $\dfrac{1}{4}$ mi
11. $\dfrac{5}{21}$
12. $1\dfrac{3}{5}$
13. $1\dfrac{1}{3}$ ft
14. 1 cup
15. $\dfrac{1}{2}$ cup
16. $2\dfrac{1}{2}$
17. $2\dfrac{1}{5}$
18. $\dfrac{7}{9}$
19. 6
20. $1\dfrac{2}{3}$
21. $2\dfrac{2}{3}$
22. $6\dfrac{1}{5}$
23. $5\dfrac{1}{7}$
24. $2\dfrac{3}{10}$
25. $2\dfrac{4}{5}$
26. $11\dfrac{1}{4}$
27. $\dfrac{1}{4}$
28. 2 hours
29. $1\dfrac{2}{5}$ miles

Pages 38–39, Adding and Subtracting Unlike Fractions

1. 4
2. 3
3. 5
4. 3
5. 6
6. 12
7. 10
8. 10
9. 25
10. 15
11. 25
12. 3
13. 40
14. 15
15. 20
16. 12
17. 30
18. 30
19. 4
20. 8
21. 21
22. 15
23. 24
24. 20
25. $\frac{4}{20} + \frac{5}{20} = \frac{9}{20}$
26. $\frac{9}{21} - \frac{7}{21} = \frac{2}{21}$
27. $\frac{4}{10} + \frac{1}{10} = \frac{1}{2}$
28. $\frac{5}{12} - \frac{3}{12} = \frac{1}{6}$
29. $\frac{3}{9} + \frac{5}{9} = \frac{8}{9}$
30. $\frac{7}{8} + \frac{4}{8} = 1\frac{3}{8}$
31. $\frac{10}{15} - \frac{4}{15} = \frac{2}{5}$
32. $\frac{5}{10} - \frac{1}{10} = \frac{2}{5}$
33. $\frac{5}{6} + \frac{2}{6} = 1\frac{1}{6}$
34. $\frac{20}{36} - \frac{9}{36} = \frac{11}{36}$
35. $\frac{9}{15} - \frac{5}{15} = \frac{4}{15}$
36. $\frac{5}{8} - \frac{2}{8} = \frac{3}{8}$

Pages 40–41, Multiplying Fractions

1. $\frac{4}{15}$
2. $\frac{5}{16}$
3. $\frac{1}{10}$
4. $8\frac{1}{3}$
5. 8
6. $\frac{1}{2}$
7. 5
8. 5
9. $\frac{3}{40}$
10. $\frac{1}{18}$
11. $\frac{1}{4}$
12. $\frac{4}{75}$
13. $\frac{8}{33}$
14. $\frac{3}{8}$ lb
15. 31 r 1
16. $\frac{8}{15}$
17. $\frac{17}{8}$
18. $\frac{13}{4}$
19. $\frac{9}{2}$
20. $\frac{27}{8}$
21. $\frac{11}{2}$
22. $\frac{49}{4}$
23. $\frac{97}{30}$
24. $\frac{26}{5}$
25. $\frac{46}{7}$
26. $\frac{38}{5}$
27. $\frac{9}{5}$
28. $\frac{7}{4}$
29. $\frac{1}{4}$

30. $5\frac{1}{2}$
31. $5\frac{5}{6}$
32. $2\frac{4}{5}$

Page 42, Canceling Before You Multiply

1. $\frac{1}{15}$
2. $\frac{1}{4}$
3. $\frac{1}{9}$
4. $\frac{1}{36}$
5. $\frac{1}{10}$
6. 96 square feet
7. $\frac{2}{5}$ oz
8. $\frac{1}{2}$ cup
9. 10

Page 43, Dividing with Fractions

1. $1\frac{1}{3}$
2. $\frac{7}{15}$
3. 4
4. $\frac{5}{56}$
5. $4\frac{2}{7}$
6. $13\frac{1}{2}$
7. 7
8. 5
9. 36
10. 6
11. 9

Page 44, Solving Mixed Word Problems

1. $\frac{1}{2}$
2. 7 volumes
3. $12\frac{1}{4}$ cups
4. 3 bushels
5. 22

6. $2\frac{3}{4}$ hr
7. $48.90
8. 64 oz
9. $\frac{13}{60}$

Pages 45–46, Fractions Skills Practice
1. B
2. H
3. D
4. F
5. E
6. J
7. C
8. G
9. C
10. G
11. D
12. G
13. A
14. K
15. B
16. H
17. D
18. F
19. D

Page 47, Positive and Negative Numbers
1. <
2. <
3. >
4. >
5. <
6. <
7. >
8. <
9. =
10. <
11. <
12. >
13. <
14. >
15. 3
16. 9
17. 6
18. 14
19. −5, −2, 0, 1

Page 48, Adding Signed Numbers
1. 9
2. −9
3. 2
4. −2
5. −8
6. 3
7. −6
8. −6
9. 0
10. −4
11. −4
12. −24
13. B
14. 2°C

Page 49, Subtracting Signed Numbers
1. 9
2. −8
3. −5
4. 0
5. 17
6. 30
7. −2
8. −10
9. 12
10. C
11. F
12. A
13. 30,500
14. B

Page 50, Multiplying and Dividing Signed Numbers
1. −25
2. 6
3. −16
4. −30
5. −32
6. 100
7. 40
8. −12
9. −120
10. −2
11. −4
12. 20
13. −3
14. 10
15. −120

16. −12
17. $\frac{1}{5}$
18. $-\frac{5}{11}$
19. $-\frac{1}{4}$
20. $\frac{2}{7}$
21. $-\frac{1}{4}$

Page 51, Solving Mixed Word Problems
1. He owed money.
2. July 14
3. They charged him money.
4. −$157.13
5. −8°F
6. $20.00
7. 50 seconds

Pages 52–53, Signed Numbers Skills Practice
1. A
2. H
3. A
4. H
5. C
6. F
7. C
8. G
9. C
10. J
11. B
12. H
13. D
14. J
15. A
16. F
17. B

Page 54, Writing Ratios
1. $150.00 : 3 *or* $50.00 : 1
2. 1,290 : 2,580 *or* 1 : 2
3. $\frac{1}{24}$
4. $\frac{18}{9}$ *or* $\frac{2}{1}$
5. $\frac{25}{5}$ *or* $\frac{5}{1}$

Pages 55–56,
Writing Proportions
[These ratios may be inverted (turned over) as long as both ratios in the proportion are inverted.]

1. $\dfrac{2 \text{ eggs}}{12 \text{ muffins}} = \dfrac{? \text{ eggs}}{30 \text{ muffins}}$
   ? = 5 eggs

2. $\dfrac{\$60.00}{20 \text{ minutes}} = \dfrac{?}{90 \text{ minutes}}$
   ? = $270

3. $\dfrac{14 \text{ ounces}}{\$2.50} = \dfrac{21 \text{ ounces}}{?}$
   ? = $3.75

4. 3
5. 9
6. 3
7. 3
8. 7,920
9. 21
10. 60.96 cm
11. 15
12. 100 minutes, or 1 hour, 40 minutes
13. $1,100
14. 25.2, or 25

Pages 57–58, Percent
1. 100
2. 130%, 650%, and 211% should be circled.
3. 50%
4. 25%
5. 0.5
6. $\frac{1}{4}$%, 50%, $50\frac{1}{2}$% 150%
7. 40
8. 0.13
9. 0.05
10. 1.15
11. 0.045
12. 0.75
13. 0.006
14. 30%
15. 2%
16. 15%
17. 190%
18. 4,020%
19. 12%

20.

Percent	Fraction	Decimal
5%	$\frac{1}{20}$	0.05
10%	$\frac{1}{10}$	0.1
15%	$\frac{3}{20}$	0.15
20%	$\frac{1}{5}$	0.2
25%	$\frac{1}{4}$	0.25
30%	$\frac{3}{10}$	0.3
$33\frac{1}{2}$%	$\frac{67}{200}$	0.335
40%	$\frac{2}{5}$	0.4
$44\frac{3}{4}$%	$\frac{179}{400}$	0.4475
50%	$\frac{1}{2}$	0.5
$57\frac{1}{4}$%	$\frac{229}{400}$	0.5725
60%	$\frac{3}{5}$	0.6
72%	$\frac{18}{25}$	0.72
75%	$\frac{3}{4}$	0.75
80%	$\frac{4}{5}$	0.8
90%	$\frac{9}{10}$	0.9

21.

$\frac{1}{3}$	33.3%	$\frac{2}{3}$	66.7%
$\frac{1}{12}$	8.3%	$\frac{8}{9}$	88.9%
$\frac{5}{7}$	71.4%	$\frac{1}{8}$	12.5%

Page 59, Finding a Percent of a Number
1. 208
2. 3.57
3. $93.62
4. 22
5. $3.00
6. $71.18
7. $28.80
8. 17,340
9. 10,547

Page 60, Finding What Percent One Number Is of Another
1. 24%
2. 4%
3. 140%
4. 56%
5. 25%
6. 10%
7. 80%
8. 16.7%

Page 61, Finding the Total When a Percent Is Given
1. 35
2. 15
3. 250
4. 208
5. $160,000
6. $21,400.00
7. 86 million

Page 62, Mixed Practice with Percent
1. 1,720
2. $600
3. 75%
4. 550%
5. $2,496
6. $296.64
7. $130.00

Pages 63–64, Ratio and Percent Skills Practice
1. A
2. J
3. E
4. F
5. B
6. G
7. A
8. G
9. C
10. J
11. B
12. F
13. B
14. H
15. C

Page 65, Reading a Table

1. Marcus
2. Adolfo
3. cooking
4. 4
5. Sunday
6. 3
7. Monday
8. Sunday

Page 66, Reading a Bar Graph

1. Accept any answer between $360,000 and $370,000.
2. Accept any answer between $8,000 and $20,000.
3. Accept any answer between $60,000 and $70,000.
4. Accept any answer between $180,000 and $195,000.

Page 67, Reading a Line Graph

1. Accept any answer between 2.6 million and 2.9 million.
2. 1940, 1945, 1955 and 1970
3. Accept any answer between 8.1 million and 8.4 million.
4. 1975 to 1995
5. 1995
6. The U.S. was fighting in World War II.

Page 68, Using Numbers in a Graph

1. $5,900,000
2. 7
3. Accept any answer from 12 to 15.
4. A
5. 159,000,000
6. Johnson Publishing

Page 69, Finding the Percent of Change

1. 43%
2. 1930 to 1940
3. Accept any answer between 28% and 36%.
4. Accept any answer between 21,000,000 and 22,000,000.
5. Accept any answer between 1.4 and 1.5.

Page 70, Finding the Mean, Median, and Mode

Median Time per doc.	Mode	Mean Time	
		per doc.	per page
8.5	None	8.5	0.7
11	10	11.5	1.0
8	None	8.6	0.7
14.5	14	15	1.3

Pages 71–72, Trends and Predictions

1. Frodo's Ice Cream, Glenn's Pot. Chips, and Health Nut Cereal
2. Arizona Chips
3. Frodo's Ice Cream
4. Health Nut Cereal
5. A
6. J
7. 1916, 1940, 1944
8. The timing became more precise.
9. the 1968 time of 9.95 seconds
10. B
11. 4
12. 72   26   32   38
13. 40

Page 73, Reading a Circle Graph

1. Cuba
2. 60%
3. $\frac{7}{10}$
4. 2.2 million
5. 13.2 million

6. B
7. G
8. 338
9. 443

Pages 74–75, Reading a Complex Table or Graph

1. 3
2. 8
3. the Saturday day shift
4. $84.00
5. $1,584.00.
6. a little over 50 years (Accept any answer between 50 and 52.)
7. Accept any answer between 77 and 79 years.
8. White Women born in 1996
9. Accept any answer between 5 and 8 years.
10. Accept any answer between 22 and 25 years.
11. women
12. Accept any answer between 67% and 70%.
13. Black Men
14. the difference between the life expectancies of Black Men and Black Women
15. C

Pages 76–77, Data Interpretation Skills Practice

1. A
2. F
3. A
4. H
5. B
6. F
7. B
8. G
9. C

Page 78, Products and Factors

1. 4^3
2. 3^2
3. 5^4

4. 2^5
5. 10^3
6. 3^4
7. $(-12)^3$
8. 0.8^2
9. 8
10. 16
11. 81
12. 100,000
13. 25
14. 1,331
15. 9
16. 2^5
17. $\frac{1}{32}$
18. 1
19. 123
20. twelve to the sixth power

Page 79, Solving Problems with Powers
1. 10
2. 19
3. 73
4. 120
5. 45
6. 180
7. 125
8. 8
9. $\frac{1}{16}$
10. 50
11. 27
12. 1
13. 49
14. 54
15. 49
16. 126
17. 34
18. 1,000
19. $\frac{1}{5}$
20. $\frac{1}{12}$
21. $\frac{4}{81}$
22. 105

Page 80, Simplifying Powers
1. 6^5
2. 5^2

3. 16^5
4. Cannot be simplified
5. 12^4
6. Accept any of $(7 \times 12)^6$ or 84^6 or "Cannot be simplified."
7. Accept any of $(\frac{14}{12})^3$ or $(\frac{7}{6})^3$ or "Cannot be simplified."
8. $\frac{1}{52^3}$
9. 6^7
10. 10^9
11. 67^9
12. 71^6
13. $\frac{1}{14^{15}}$
14. 8^4

Pages 81–82, Scientific Notation
1. 3,050,000
2. 410,000,000,000
3. 20,035,000,000
4. 0.000045
5. 910,500
6. 0.0731
7. 0.000000008
8. 90,000
9. 5.6×10^9
10. 3.4×10^6
11. 5.05×10^{11}
12. 1.92×10^{-8}
13. 6.071×10^{13}
14. 4.006×10^{-8}
15. 5.1698×10^{-14}
16. 1.9×10^9
17. 5.4×10^6
18. 1.002×10^1
19. 8.7×10^{-13}
20. 9.1×10^{-6}
21. 3.012×10^6
22. 8.1117×10^{-12}
23. 9.9×10^{-6}
24. 7.0102×10^8
25. 0.01505
26. 6,200
27. 70.9
28. $\frac{1}{0.05}$ or 20
29. 0.0061×100 or 0.61
30. 4,000 people

Page 83, Square Root
1. 3
2. 4
3. 11
4. 6
5. 10
6. 15
7. 8
8. 14
9. 2 and 3
10. 3 and 4
11. 4 and 5
12. C
13. H
14. 5
15. 1
16. 8
17. $\frac{3}{5}$

Pages 84–85, Exponents, Powers, and Roots Skills Practice
1. B
2. F
3. D
4. J
5. C
6. H
7. B
8. G
9. C
10. H
11. A
12. F
13. B
14. F
15. B
16. J
17. C
18. H
19. B
20. F
21. A
22. F
23. D

Page 86, Patterns
1. 15
2. The drawing should have stacks of 5, 4, 3, 2, and 1 brick(s).
3. 2, 4
4. B
5. H
6. 22
7. D
8. 300; divide by 3
9. 150

Page 87, Completing Number Sentences
1. ÷
2. −
3. ×
4. ×
5. ×
6. −
7. − or +
8. ÷
9. −
10. +
11. ×
12. +
13. ×
14. ÷
15. ×
16. +
17. −
18. ÷
19. −
20. +
21. +
22. ÷
23. −
24. ×
25. ÷
26. ×
27. ×

Page 88, Inverse Operations
1. 94
2. 216
3. 24
4. $\frac{4}{25}$
5. 1.27

6. −28
7. −17
8. 58
9. 155
10. 132
11. 2
12. $\frac{7}{12}$
13. 12
14. 15
15. 7
16. 2.02
17. $4\frac{7}{10}$
18. 6
19. 0.03
20. 5
21. 7
22. $\frac{1}{6}$
23. 12

Page 89, Writing Letters and Symbols for Words
1. $\frac{x}{14}$
2. $x + 2$ (or $2 + x$)
3. $54x$
4. x^2
5. $x - 63$
6. $x + 42$ (or $42 + x$)
7. $\frac{x}{2}$ or $(\frac{1}{2})x$
8. $12x$
9. $2x$
10. $100 - x$ (or $x - 100$)
11. $-2 - x$
12. $\frac{x}{13}$
13. $\frac{-4}{x}$
14. $\frac{1}{\sqrt{x}}$
15. $\frac{2x}{5}$ or $\frac{2}{5}x$

Pages 90–91, Writing Longer Algebraic Expressions
1. A
2. G
3. A
4. H

5. B
6. F
7. A
8. H
9. B
10. G
11. $4 + 2n$ or $2n + 4$
12. $\sqrt{n} - 5$
13. $\frac{n}{2} + 30$ or $30 + \frac{n}{2}$
14. $4(n^2 - 2)$
15. $3(4 + n) - 12$ or $3(n + 4) - 12$
16. $\frac{n-5}{n}$
17. $-\frac{1}{3}(n + 5)$ or $-\frac{1}{3}(5 + n)$
18. $8 + \frac{n}{5}$ or $\frac{n}{5} + 8$
19. $60 - 3n$
20. $7n - 10$
21. $\sqrt{n} - 12$
22. $\frac{n-6}{12}$

Page 92, Writing Algebraic Equations
(The answers below just show one likely way to set up each problem.)
1. $b + 2b = \$72,000$
2. $\$159 + \$45h = \$675$
3. $m + \frac{m}{2} = \$450$
4. $75(4) + 65(2) = d$
5. $\$25.00 + (c - 40)(\$0.04) = b$
6. $25L = L + \$300,000$
7. $9(t + 1) = 15t$

Pages 93–94, Solving an Equation with an Addition or Subtraction Sign
1. $x = 175$
2. $x = 492$
3. $x = 118$
4. $x = 8$
5. $x = 39$
6. $x = -4$
7. $x = 12.7$
8. $x = 19$
9. $x = 16$

10. $x = 82$
11. $x = 4.86$
12. $x = 29$
13. $x = 12$
14. $x = 23$
15. $x = 5$
16. $x = 14$
17. $x = 49$
18. $x = 32$
19. $x = 3$

Page 95, Solving an Equation with a Multiplication or Division Sign

1. $n = 2$
2. $x = 21$
3. $a = -5$
4. $n = 216$
5. $c = 144$
6. $a = 8$
7. $n = 5.1$
8. $x = 70$
9. $p = 750$
10. $x = \frac{6}{7}$
11. $x = 5$
12. $n = 10$
13. $b = 30$
14. $t = 32$
15. $c = 10$
16. $s = 12$
17. $v = 20$
18. $x = 16$

Page 96, Combining Terms

1. $7ab$
2. cannot be combined
3. $12x^2$
4. $-6a^2$
5. cannot be combined
6. $16an$
7. $10b^7$
8. 2
9. x
10. $4t$
11. cannot be combined
12. xy
13. $3a^3b$
14. $8u$
15. $4y^2$

Page 97, Solving an Equation with Parentheses

1. $x = 100$
2. $x = 15$
3. $b = 5$
4. $n = 7$
5. $v = 3$
6. $y = 50$
7. $c = 1$
8. $d = 100$
9. $n = 20$
10. $x = 3$
11. $e = 6$
12. $f = -5$
13. $a = 2$
14. $b = -\frac{2}{3}$
15. $a = 2$
16. $3x^2 + 15x$
17. $2a^2b + 2ab^2$
18. $44c - 4c^3$
19. $2xy^2 + y^4$
20. $4a^2b + 4ab^2$
21. $6x^3 + 6x^4$

Page 98, When a Sum or Difference Appears in a Fraction

1. $x = 10$
2. $x = 8$
3. $x = 10$
4. $x = 6$
5. $x = 3$
6. $x = -1$
7. $3 - x$
8. $5 + \frac{1}{x}$
9. $2 - x$
10. $1 - 2x$
11. $1 + x + 14$ (or $x + 15$)
12. $3x + 9$

Page 99, When a Variable Appears Twice in an Equation

1. $y = 21$
2. $a = 3$
3. $x = 12$
4. $f = 0.5$
5. $r = 5$
6. $n = 4$

7. $t = 21$
8. $n = -13$
9. $x = 4$
10. $a = 6$
11. $b = 32$
12. $n = 2$

Page 100, Solving Inequalities

1. $a > 27$
2. $b < -25$
3. $x > 3\frac{3}{4}$
4. $u \geq 4$
5. $y < 2$
6. $n \leq 16$
7. $a < 205$
8. $y < 54$
9. $x < 4$ (or $x > -4$)
10. $a > 7$
11. $y \leq 3\frac{1}{3}$
12. $b < 2$ (or $b > -2$)

Page 101, Using Formulas

1. 5 years
2. 20 feet
3. 3.7 hours
4. $1,456.00

Page 102, Solving Word Problems Using Algebra

1. D
2. 300 women
3. 13
4. $14,000,000
5. $48,000
6. $n = 77$
7. $80.25
8. 55

Pages 103–104, Algebra Skills Practice

1. C
2. H
3. A
4. K
5. B
6. J
7. C

Answer Key

8. G
9. C
10. H
11. B
12. H
13. C
14. F
15. B
16. J
17. D
18. G
19. C

Page 105, Reading a Scale
1. 18
2. 66 or 67
3. 64
4. $1\frac{3}{8}$ cups
5. 65
6. Accept 31 or 32.
7. 30°
8. 10
9. 2
10. 55

Page 106, Standard Units of Measure
1. B
2. G
3. H
4. B
5. 1 foot
6. 5 feet
7. 1 pound
8. 10 ounces
9. 1 pint
10. 1 gallon
11. 1 gallon

Pages 107–108, Using a Standard Ruler
1. $2\frac{1}{4}$ in.
2. $1\frac{3}{4}$ in.
3. $\frac{1}{2}$ in.
4. 2 in.
5. $\frac{1}{4}$ in.

6. $1\frac{1}{8}$ in.
7. $2\frac{3}{8}$ in.
8. $2\frac{5}{8}$ in.
9. $\frac{3}{8}$ in.
10. $\frac{1}{2}$ in.
11. $\frac{5}{16}$ in.
12. $\frac{3}{8}$ in.
13. $\frac{7}{16}$ in.
14. $1\frac{1}{16}$ in.
15. $\frac{1}{4}$ in.
16. $1\frac{3}{16}$ in.
17–30. Students' answers may vary.

Page 109, Converting Within the Standard System
1. 3
2. 30
3. 4.5 *or* $4\frac{1}{2}$
4. 3
5. 4
6. 9
7. $1\frac{1}{2}$
8. 19
9. 70
10. 11
11. 147
12. 6
13. 1, 3
14. 2, 3
15. 12
16. 2
17. 128
18. 22,704
19. 2
20. $\frac{7}{8}$

Page 110, The Metric System
1. gram
2. milliliters
3. kilogram
4. centimeters
5. grams
6. 500
7. 20
8. 4.2
9. 250
10. 0.5
11. 1,000
12. 750
13. 15
14. 2,150
15. 12,700
16. 250
17. 5,300

Page 111, Using a Metric Ruler
1. 6 cm
2. 3 cm
3. 1 cm
4. 4 cm, 5 mm
5. 5 cm, 5 mm
6. 6 cm, 3 mm
7–16. Students' answers will vary.

Page 112, Comparing Standard and Metric Units
1. 1 meter
2. 1 meter
3. 1 gallon
4. 1 mile
5. 1 ounce
6. 1 inch
7. 1 kiloliter
8. 1 liter
9. 1 kilogram
10. 16.36
11. 0.82
12. 20
13. 9.09
14. 1.82
15. 3.64
16. 9.8

Page 113, Adding and Subtracting Mixed Measurements

1. 2 hr, 30 min *or* $2\frac{1}{2}$ hr

2. 3 ft, 3 in. *or* $3\frac{1}{4}$ ft

3. 14 yd
4. 6 gal, 2 qt *or* $6\frac{1}{2}$ gal
5. 6 hr, 10 min
6. 16 lb, 5 oz
7. 30 yd
8. 25 ft, 5 in.
9. 4 hr, 15 min *or* $4\frac{1}{4}$ hr
10. 5 qt, 1 cup
11. 2 lb, 6 oz
12. 7 yd, 1 ft
13. 1 ft, 10 in.

14. 4 gal, 2 qt *or* $4\frac{1}{2}$ gal

15. 2 hr, 45 min *or* $2\frac{3}{4}$ hr
16. 1 cup

Page 114, Multiplying and Dividing Mixed Units

1. 174 ft
2. 52 lb 8 oz *or* $52\frac{1}{2}$ lb
3. 34 yd
4. 3 in.
5. 10 oz
6. 6
7. 132
8. 75 ft
9. 6
10. 5

Page 115, Finding Perimeter

1. 64 ft, 4 in.
2. 55 ft, 10 in.
3. 78 cm
4. 80 in.
5. 9 m
6. 36 mm
7. 13 ft
8. 13 mi
9. 20 m
10. 2 feet

Page 116, Finding the Circumference of a Circle
(Answers will vary slightly as shown depending upon the value of π used.)

1. Accept 12.56 or 12.57 inches.
2. cannot tell
3. 62.8 or 62.9 cm
4. 18.84 or 18.85 mm
5. 31.4 or 31.43 inches
6. 3.93 or 3.94 feet
7. 25.12 or 25.14 feet
8. 7.85 or 7.86 feet

Page 117, Finding Area

1. about 7 ft^2
2. 96 ft^2
3. 71 ft^2
4. 160 cm^2
5. 35 cm^2
6. 390 cm^2

Page 118, Finding Volume

1. 25 in.^{3}
2. 0.25 ft^3
3. 56.52 in.^{3}
4. 17.7 m^3
5. 8 cm^3
6. about 5
7. 2.6 ft^3

Page 119, Changes in Time

1. 5:55 A.M.
2. 8:17 A.M.
3. 1 hour, 53 minutes
4. 9:30 P.M.
5. 2:00 P.M.
6. 7:52
7. 50

Pages 120–121, Measurement Skills Practice

1. A
2. J
3. C
4. F
5. C
6. J
7. A
8. F
9. A
10. J
11. B
12. G
13. B
14. H
15. C
16. G

Page 122, Visualizing

1. C
2. J
3. 6
4. 12
5. A

Page 123, Basic Concepts

1. .E

2. C ———— B

3. ← V —— W →

4.

5. angle
6. line
7. B
8. Two planes intersect in a line.

Page 124, Lines

1. line EF (or line FE)
2. line AD and line BG *or* line FE and line BG
3. intersecting
4. 2nd Street
5. Each "Avenue" is perpendicular to each "Street."
6. segments (of parallel lines)

Page 125, Angles

1. B
2. G
3. A
4. G

5. B
6. H
7. C
8. G

Page 126, Types of Angles
1. acute
2. right
3. obtuse
4. right
5. acute
6. obtuse
7. right
8. obtuse
9. acute
10. obtuse
11. A: obtuse; B: acute; C: acute; D: right
12. 360°
13. G

Pages 127–128, Related Angles
1. 20°
2. 155°
3. 50°
4. 50°
5. 150°, 30°, 150°
6. A
7. ∠ADE or ∠EDA
8. 55°
9. 125°
10. 90°
11. H
12. ∠DBA, ∠ABD, ∠BDC, or ∠CDB
13. C
14. H

15. B
16. ∠u

Page 129, Polygons
1. the swings
2. the water sprinkler
3. trapezoid
4. pentagon
5. the merry-go-round
6. the merry-go-round; regular octagon

Page 130, Types of Triangles
1. isosceles, 20°
2. scalene, 25°
3. scalene (and right), 90°
4. equilateral, 60°

Page 131, Right Triangles
1. none
2. E and F
3. J and H
4. $\sqrt{18}$ inches
5. $\sqrt{41}$ units
6. $\sqrt{32}$ units
7. $\sqrt{24.25}$ ft

Page 132, Congruent Figures and Similar Figures
1. triangle B
2. triangle C and triangle E (Triangle E is also congruent to triangle F.)

3. 55°
4. side *GH*
5. No. Corresponding angles are not equal.
6. side *KH*
7. ∠*HJK*
8. side *JK*

Pages 133–134, Solving Problems with Similar Triangles
1. 14 miles
2. ∠*GEF*
3. 90°
4. side *XY*
5. 30°
6. 20 ft
7. 60°
8. side *XU*
9. 1 foot
10. A
11. H
12. 10
13. C
14. 18 ft

Pages 135–136, Geometry Skills Practice
1. C
2. F
3. C
4. J
5. D
6. G

Glossary of Common Terms

absolute value (| |): a number's distance from zero. Examples: |5| and |−5| are both 5. One is five more than zero; the other is five less.

algebraic expression: one or more variables and one or more numbers put together with an operation like addition. Example: $3x + 5$

average or mean: the sum of all the values in a set divided by the total number of values. Example: To find the average of the set 1, 5, 4, 6, 5, and 3, add all six numbers together. Then, divide the sum (24) by 6. The average is 4.

canceling: simplifying a fraction multiplication problem. To cancel, use the same number to divide one of the numerators and one of the denominators in the problems.

circumference: the distance around a circle

complementary angles: angles whose sum is 90°

decimal: a number containing digits to the right of the ones place. These digits represent part of one. A small dot, called the decimal point, separates whole number digits from decimal digits Example: In $13.42, digits to the left of the decimal point represent 13 whole dollars. Digits to the right represent 42 hundredths of a dollar.

denominator: the bottom number in a fraction. The denominator tells how many parts were in the whole object or how much was in the whole amount. The **least common denominator** of two fractions is the smallest number that can be used as a denominator for both. Example: the least common denominator of $\frac{1}{3}$ and $\frac{2}{5}$ is 15.

diameter: the distance across a circle at its widest point (through its center)

dividend: the number in a division problem that is being divided. Example: In the division problem $120 \div 6 = 20$, 120 is the dividend.

divisor: the number in a division problem that you are dividing by. Example: In the division problem $120 \div 6 = 20$, 6 is the divisor.

equivalent: having the same value. Examples: $\frac{1}{2}$, $\frac{4}{8}$, and $\frac{5}{10}$ are all equivalent fractions.

equilateral triangle: a triangle with three equal sides and three equal angles

exponent: a number that shows how many times the base is used as a factor in a product. Example: In the expression 8^5, 5 is the exponent.

formula: a rule given in algebraic form. Example: $i = prt$

hypotenuse: the longest side in a right triangle; the side opposite the right angle

inequality: a statement that two expressions are not equal. Example: $2a\ b < 42$

intersecting lines: lines that cross

inverse operations: operations that undo each other. Addition and subtraction are inverse operations. So are multiplication and division.

isosceles triangle: a triangle with at least two equal sides and at least two equal angles

lowest or simplest terms: the form of a fraction or ratio that cannot be reduced. In other words, there is no whole number, other than 1, that can evenly divide both the numerator and denominator. Example: $\frac{6}{8}$ is not in simplest terms because both 6 and 8 can be evenly divided by 2.

median: the middle value in an ordered list of numbers, or the number halfway between the two middle values.

mode: the number in a set that appears most often

multiples of a number: numbers that can be evenly divided by that number. Example: 9, 15, 21, and 30 are all multiples of 3. A **common multiple** of two numbers is a number that can be evenly divided by both.

negative number: a number with a value less than zero. Negative numbers are always shown with a negative sign (–).

numerator: the top number in a fraction. The numerator tells how many parts the fraction represents.

parallel lines: two lines that are always the same distance apart. Parallel lines never cross.

percent (%): a fraction with an unwritten denominator of 100. The word percent means "per hundred." Examples: 20 out of 100 is $\frac{20}{100}$ or 20%.

perpendicular lines: two lines that form right angles when they meet

pi (π): approximately 3.14 or $3\frac{1}{7}$

place: the location of a digit in a number. Example: 173 has three places. The digit 1 is in the hundreds place, 7 is in the tens place, and 3 is in the ones place.

polygon: a flat shape that has straight sides. Examples: rectangle, hexagon, pentagon. In a **regular polygon,** like a square, all sides have the same length.

power: an expression containing a base and an exponent. Example: 5^7

proportion: an equation showing that one ratio equals another. Example: $\frac{3}{4} = \frac{6}{8}$

radius: the distance from any point on a circle to its center. The radius of a circle is always half its diameter.

ratio: a comparison of two numbers, such as Ron has 3 days off for every 7 days he works. Ratios can be written in three ways: 3 to 7, 3 : 7, or $\frac{3}{7}$.

reciprocal of a number: one divided by that number. The reciprocal of a fraction is that fraction turned "upside down." Examples: 7 and $\frac{1}{7}$ are reciprocals and $\frac{3}{8}$ and $\frac{8}{3}$ are reciprocals.

right angle: an angle that measures 90°. Each corner in a square or a rectangle is a right angle.

scientific notation: a method of writing numbers using powers of ten. Example: 5,400,000,000 written in scientific notation is 5.4 10^9.

similar figures: figures with the same shape, but not necessarily the same size

square root of a number: a number whose square equals that number. Example: 5 is the square root of 25.

supplementary angles: angles whose sum is 180 °

term: each of the quantities joined with addition or subtraction signs in an equation. Example: The expression $2ab^2 + \frac{a}{3}b$ contains two terms: $2ab^2$ and $\frac{a}{3}b$.

unknown or variable: a letter (or variable) in an algebra expression. Unknowns take the place of numbers that have not been identified. Example: In the expression $2a + 3 = 11$, a is the unknown.

vertical angles: opposite angles formed by a pair of intersecting lines

Glossary of Common Terms

Index

Multiplying
  canceling, 42
  carrying, 6–7
  checking, 6
  cross multiplying, 56
  decimals, 24–25
  fractions, 33, 40–42
  inverse of, 6, 88, 95
  measurements, 107, 111
  signal words for, 11, 25, 33
  signed numbers, 50
  whole numbers, 6–7, 40

N
Negative numbers. *See* Signed
  numbers.
Number line, 47–49
Number order, 4, 8, 14

O
Octagons, 129
One, properties of, 87, 98
Order of operations, 99

P
Parentheses, 79, 87, 89–90,
  97, 99
Patterns, 72, 86
Percent, 57–62, 69, 73, 75
Perimeter, 112
Pi (π), 113–115
Place value, 1–2, 16, 20–21
Planes/Plane figures , 112–114,
  120, 125–131
Points, 120
Polygons. *See* Plane figures.
Positive numbers. *See* Signed
  Numbers.
Powers, 24, 78–82, 94, 108
Predictions, 71–72, 75
Prime numbers, 34
Proportions, 55–56, 107, 109,
  130–131
Pythagorean Theorem, 128

R
Radius, 114–115
Rates, 54
Ratios, 54–56, 68
Reciprocals, 43
Regrouping *See* Borrowing
  and Carrying

Remainders, 8, 10, 28, 35
Rounding, 16–17, 20, 22, 25–26,
  28–29, 58–59, 60–61, 69

S
Scales, 105
Scientific notation, 24, 81–82
Signal words, 11, 15, 17, 25, 33,
  54
Signed numbers, 47–51, 79
Similarity, 129–131
Simplest terms, 34–35, 50
Simplifying
  equations, 96–99
  fractions. *See* Fractions,
    reducing.
  measurement units, 110–111
  problems,
    with division, 27, 32–33, 43
    with fractions, 42–43
    with powers, 80
    with signs, 48–50
Solids, 115, 119
Square roots, 83, 128
Standard Units of Measurement,
  106–107, 109
Subtracting
  borrowing, 4–5, 37, 110
  checking, 4, 36–37
  decimals, 23
  fractions, 36–39
  inverse of, 4, 23, 36–37, 49, 88,
    93–94
  measurements, 110
  on the number line, 49
  signal words for, 11
  signed numbers, 49
  whole numbers, 4–5, 68
  from zeros, 5, 23

T
Tables, 65, 70, 72, 74
Terms, 94, 96, 99
Time, 105–107, 110–111, 116
Trends, 67, 71–72
Triangles, 126–131

U
Unknowns, 89

V
Variables *See* Unknowns
Volume, 115

W
Whole numbers
  adding, 3
  comparing, 2, 35
  decimal forms, 22
  dividing, 8–10, 28
  fraction forms, 35, 40, 43
  multiplying, 6–7, 40
  naming, 2
  place value in, 1–2
  subtracting, 4–5
  word problems, 11–15, 17, 68,
    70, 74–75
Word problems, with
  decimals, 23, 25, 27, 29
  estimation, 15, 17
  fractions, 32–33, 36–37, 39–40,
    42–44
  geometric figures, 114–116,
    123, 128, 130–131
  measurement/perimeter/area/
    volume, 107, 111–116
  percent, 57, 59–62, 69, 73
  powers, 82
  ratios/proportions, 54–56
  rounding, 17
  set up, 11–12, 14, 55–56
  signal words, 11, 15, 17, 25,
    33, 54
  signed numbers, 48–49, 51
  solving with algebra, 89–92,
    101–102
  tables or graphs, 68–70, 73–75
  time, 116
  two-steps, 12, 14, 17, 29, 51
  unnecessary information, 13
  whole numbers, 11–15, 17

Z
Zero
  as a place holder, 2, 4, 6, 8,
    22–26, 28, 59
  properties of, 5, 23, 36, 78,
    87, 98